Dragonflies and Damselflies of Costa Rica

Dragonflies and Damselflies of Costa Rica

A FIELD GUIDE

DENNIS PAULSON
WILLIAM HABER

Antlion Media
A Zona Tropical Publication
FROM
Comstock Publishing Associates
an imprint of
Cornell University Press
Ithaca and London

First published 2021 by Cornell University Press

Printed in China

Library of Congress Cataloging-in-Publication Data

Names: Paulson, Dennis, 1937– author. | Haber, William A., author.
Title: Dragonflies and damselflies of Costa Rica : a field guide / Dennis Paulson, William Haber.
Description: Ithaca [New York] : Comstock Publishing Associates, an imprint of Cornell University Press, 2021. | "A Zona Tropical publication." | Includes bibliographical references and index.
Identifiers: LCCN 2020042512 | ISBN 9781501713163 (paperback)
Subjects: LCSH: Dragonflies—Costa Rica—Identification. | Damselflies—Costa Rica—Identification.
Classification: LCC QL520.22.C67 P38 2021 | DDC 595.7/33—dc23
LC record available at https://lccn.loc.gov/2020042512

Zona Tropical Press ISBN 978-1-949469-35-6

Book design: Gabriela Wattson

For the two Philips: Philip P. Calvert, who wrote the
Odonata section of *Biologia Centrali-Americana*,
spent a year in Costa Rica, and contributed so much
to our knowledge of its dragonflies; and Philip S.
Corbet, who, through his *Dragonflies: Behavior
and Ecology of Odonata*, taught us all about this
wonderful group of insects

Contents

Preface

As one of the most popular ecotourism destinations in the world, Costa Rica is long overdue for a field guide to its dragonflies and damselflies (Odonata). Like butterflies, they are large, colorful, active, and diurnal, and thus of great interest to even casual naturalists. Short of visiting museum collections or poring through the taxonomic literature, there has been no way for people to put accurate names on what they were seeing, though this has slowly changed.

Carlos Esquivel started the ball rolling with his 2006 book on the Odonata of Middle America and the Caribbean. This is a fine book, full of interesting facts about the species it covers, but it contains only a subset of the species that occur in this large area, with a mere 72 Costa Rican species featured and illustrated by photos. Another reference for the region is the identification guide for Central America by Steffen Förster, first published in 1999, with a second edition in 2001. This book is helpful, but only for specimens in the hand as it is a series of keys to all the genera and species, illustrated with line drawings of the fine points needed for such a task.

No other books are available to help with the identification of Odonata in Central America. There are finally books for South America, albeit for regions that are mostly at some distance from Costa Rica. These cover the Yungas of Argentina by Natalia von Ellenrieder and Rosser Garrison (2007); the Serra dos Orgaos in southern Brazil by Tom Kompier (2015); Trinidad and Tobago by John Michalski (2015); and the Colombian Cordillera Occidental by Cornelio Bota-Sierra et al. (2019).

Just a few years ago, we started talking about the need for this book. Paulson made extensive collections during more than a year spent in Costa Rica in 1966–1967, and he has continued to make visits in the ensuing years. Haber has lived in Costa Rica and studied the odonate fauna for the past 16 years. Fortunately, when we first considered writing this book, we already had large collections of specimens and many photos and scans of Costa Rican species.

When the last paper on Costa Rican dragonfly diversity was published in 2000 by Alonso Ramírez, Carlos Esquivel, and Dennis Paulson, it listed 268 species for the country. We are now up to 283 species and counting. As this preface was being typed, a graduate student in Costa Rica, Jareth Román, sent us a photo of a dragonfly she had just caught—a Fairy Skimmer (*Elga leptostyla*), a new record for the country—from a part of the country extensively visited. At least four species of shadowdamsels (*Palaemnema*) and one species of dancer (*Argia*) have not been described and are not discussed in this book. Clearly much more remains to be discovered, and we anticipate a future edition of this book substantially different from the first.

Acknowledgments

I am deeply grateful to Gordon Orians, who took me to Costa Rica on a postdoctoral fellowship so long ago and was thus involved in planting the seeds that grew into my longtime interest in the odonates of this biodiverse country. Special thanks to Lynn Erckmann, who accompanied me on that first trip and helped me survey the odonates. I should also thank Leslie Holdridge posthumously for letting us stay at his house at La Selva, where I realized how amazingly diverse and interesting was the Costa Rican fauna. Much later, I benefitted greatly from the dragonfly nature tours organized by Steve Bird and David Smallshire and enjoyed the companionship of fellow tour-goers as we made dragonfly discoveries. I also want to thank Bill Haber for being such a great colleague and partner in putting together this book. Of course, we will always be thankful of our many colleagues who donated photos to the cause, whether we used them or not. And finally, I cannot thank enough my wife, Netta Smith, who has accompanied me on all recent trips and provides ongoing support, encouragement, and yes, love, as I continue to learn and teach about nature. We both express our deepest gratitude to John McCuen, Gabriela Wattson, and Zona Tropical for doing such a fine job of editing, designing, and producing this book and having patience with us as we struggled to put together all of its many parts.

Dennis Paulson

Many friends and colleagues helped lead me here. First among them is David Wagner, who introduced me to some of the more dazzling and elusive Costa Rican damselflies; his enthusiasm and mentoring awakened the passion that I still have for these magnificent little creatures. Dennis Paulson was also most generous with his time, encouraging my interest through e-mail, identifying images and specimens, providing scientific papers, and, later, sharing hours at watery viewing sites. Others, whose many days in the field with me are much appreciated, include Eladio Cruz, Fred Morrison, Lucas Ramírez, Fred Sibley, and Ronald Vargas, all fine naturalists, who showed me rare species and other amazing things. Rosser Garrison and Natalia von Ellenrieder enthusiastically provided identifications, technical literature, specimens, and consultation. Finally, thanks to Willow Zuchowski for unstinting support for my aberrant odonate pursuits.

William Haber

Introduction

Just as birds have their devotees, so too do dragonflies and damselflies. They are fantastic fliers—watch one go straight up out of sight to chase another dragonfly. Their behavior has been compared to that of birds—males of many species defend territories with aggressive displays, chases, and fights among them, and some species exhibit complex courtship behavior. The adults are the raptors of the insect world, feeding on other insects up to the size of dragonflies and butterflies, and their predatory behavior can be watched up close. And they really do come in all colors of the rainbow. Common Costa Rican species are black, white, gray, brown, red, orange, yellow, blue, green, and violet, in many shades and combinations, including iridescent and metallic variations.

Dragonflies are easy to observe at freshwater wetlands. They are among the largest of all insects, are conspicuous and often brightly colored, and are active during the day. In some circles, the term *dragonfly* includes damselflies, but we have chosen to differentiate them, as they are in two different suborders of Odonata. It's simplest just to use the term *odonates* when talking about them collectively. Damselflies are generally the smaller odonates, with slender abdomens, fore- and hindwings of the same shape, and relatively small eyes separated by a distance greater than the diameter of the eye, with heads that are shaped like barbells. Dragonflies are usually larger and more robust, can have broad abdomens, have hindwings conspicuously broader than the forewings, and large eyes that are close together or touching. Except for one species in the genus *Zenithoptera*, all Costa Rican dragonflies perch with the wings outspread. Most damselflies rest with them closed, although members of several Costa Rican families (Lestidae, Heteragrionidae, Philogeniidae) hold their wings open. And note that the odonates with the longest wings and bodies are the helicopter damselflies.

Odonate Biology

Odonate larvae, also called nymphs and, less often, naiads, are predators, feeding on insects, crustaceans, and other invertebrates, as well as vertebrates such as tadpoles and small fish. They capture their prey with their labium, a unique apparatus that is essentially an extensible lower lip with an enveloping basket or hooks on the end. In some fast-growing species, they emerge from their larval state after only a month in warm pond water (e.g., Red-tailed Swampdamsel and Spot-winged Glider), while large dragonflies that develop in cold water (e.g., Apache Spiketail) may take multiple years.

Large dragonflies such as darners emerge during the night, but in Costa Rica smaller clubtails and some skimmers—along with many damselflies—emerge during the day, when they can sometimes be found by a careful observer. Pond damsels undergo emergence fairly rapidly, and you can sit by the waterside to watch the entire process. Some people choose to take larvae into captivity and feed them, watching their development over a period of weeks or months.

The larva climbs from the water and generally fixes itself on a vertical substrate; clubtails and pond damsels, however, often emerge on horizontal surfaces such as sandy beaches, rocks, lily pads, and algal mats. After a short time, the thorax splits along its dorsal line, and the adult begins to push out. It frees its head and legs, rests until its legs harden, then pulls its abdomen out of the cast skin (*exuvia*, plural *exuviae*). It then pumps fluid into its wings, which expand slowly. After the wings are fully expanded, fluid is pumped back out of them into the abdomen, which expands to its full length. The adult rests in this position for a while, then opens its wings (in species that normally hold their wings open). After a further period of hardening, the freshly emerged (*teneral*) adult flutters up from its perch and flies away. Emerging larvae are subject to high mortality, often falling victim to predation or bad weather.

After emergence, tenerals quickly harden and develop their distinctive color pattern. These immature adults stay away from the water for a period that varies from a few days to as long as an entire dry season in some species. They feed actively during this time, capturing prey with their long, spiny legs and transferring it to their well-developed mandibles. They feed either in flight or after returning to a perch. *Fliers* feed like swallows and swifts; most darners, emeralds, gliders, saddlebags, clubskimmers, and sylphs forage in flight and eat their small prey while flying, usually discarding the wings. *Perchers* feed in two different ways. Clubtails, most skimmers, shadowdamsels, damselflies that rest with open wings, and dancers (a type of pond damsel) are *salliers*, watching for flying prey while perched and sallying into the air to intercept it, returning to the perch much like a flycatcher. *Hover-gleaners*, including most pond damsels (from tiny threadtails to huge helicopters), fly through the vegetation slowly looking for stationary prey, which they dart at and grab, then land to eat.

Odonate body temperature is largely controlled by air temperature—they become inactive when it cools down. Thus, at higher elevations in the tropics, with lower temperatures, odonates may remain inactive for long periods. But even on a typical day in the hot lowlands, many of them are inactive when it is cloudy, probably because their vision is most effective when they are in the sun, and poorer vision may make them less able to capture prey. When the sun comes out, they show up at the water as if by magic. Even a slight lowering of body temperature can make them less able to compete for mates. Exceptions are some of the large species, especially darners, which can elevate their body temperature by wing-fluttering in the same way that bees and butterflies do. Early in the morning, many individuals take up position on the sunny side of vegetation, warming up in that way, and they are often able to remain active late into the afternoon by similarly perching in sunny sites.

Before they are fully developed, young odonates, or *immatures*, live away from water. They maintain immature coloration, in which males and females often look alike and must be distinguished by shape; this dull coloration also serves as camouflage. As odonates develop sexually, they change to their mature coloration, their gonads develop, and they return to the

Turquoise-tipped Darner (*Rhionaeschna psilus*) is a typical dragonfly, with large eyes touching each other and broad wings held open. The forewings and hindwings are shaped differently.

Cerulean Dancer (*Argia anceps*) is a typical damselfly, with small eyes at some distance from one another. The forewings and hindwings are of the same shape and held over the abdomen.

water to breed. While at the water, males spend much of their time watching for females. Males of many species, including the larger damselflies and most skimmers, defend fixed territories near good oviposition sites, a behavior that allows them to claim exclusivity to visiting females. Many pond damsels and darners, by contrast, fly around wetlands actively searching for females, which may already be there laying eggs.

Females come to the water only to mate and lay eggs and are immediately grasped by any male of their species that sees them. The male dragonfly clasps the female by the rear margin of the head with three terminal appendages, the male damselfly by the thorax (both the prothorax and the front end of the pterothorax) with two pairs of terminal appendages. This is the *tandem* position. Before tandem is accomplished in most odonates, the male moves sperm from his eighth abdominal segment to the functional penis (independently evolved *genital ligula* in damselflies and *vesica spermalis* in dragonflies) in his second abdominal segment, where it is stored in a seminal vesicle. In most damselflies, this act is performed during tandem. The female then curls her abdomen up, with help from the male, and hooks up to that second segment, where copulation takes place, either in flight or at rest. This is the *wheel* position, unique to Odonata, and it may last only a few seconds in flight (many skimmers) or go on for much longer periods at rest.

In many damselflies and a few skimmers, the pair stays in tandem and begins to prospect for egg-laying sites, but in most other odonates they separate and the female oviposits alone, in some species followed closely by the male, who attempts to protect her from the attentions of other males. If another male grabs her and copulates, he will displace the sperm deposited in her sperm-storage organs by the previous male with specialized structures on his penis and replace it with his own. Thus, mate guarding is a critical aspect of reproductive behavior.

Most females oviposit either in plant material by means of a sharp-bladed ovipositor (all damselflies and darners) or directly into the water by tapping the surface and releasing bunches of eggs (most other dragonflies). A few species, such as spiketails and some

A mating pair of Red Saddlebags (*Tramea onusta*).

darners and clubtails, lay eggs in mud at the water's edge. The complete clutch of a female can consist of many hundreds of eggs, and she will lay them throughout her life. One mating should furnish enough sperm, stored in her sperm-storage organs, to fertilize all those eggs, although individuals of both sexes commonly mate more than once.

When the odonate egg hatches, what emerges is a distinctive stage called a *prolarva* that is quite unlike the succeeding larval stages. It does not feed, but is somewhat motile and, with rain or rising water, can help move itself from a dry oviposition site into the water, where it quickly sheds its cuticle to transition into the normal larval form. The prolarva is unknown in any other insects. The larva cannot increase in size because of its rigid exoskeleton, so to reach full size it must molt repeatedly into subsequent stages. Each stage is called an instar, and the larva passes through 11–13 of these instars prior to the final molt into an adult.

As the sex ratio at the water is usually strongly biased in favor of males, females are harassed constantly while trying to lay their eggs. They have evolved a number of ways to fend off males. In many cases, the female rejects the male by curling her abdomen under her. She may even attempt to bite him. Females often come to the water to oviposit early in the morning or late in the afternoon, when male density is low. Flying away from the water when harassed is a common behavior, and some females just land or drop into the vegetation, apparently decreasing their visibility when stationary.

Odonates in the Tropics

Biodiversity. It is safe to say that in the tropics most animal and plant groups contain more species than in temperate regions. This is dramatically true for some groups, less so for others. Odonates represent a group in which species diversity increases at lower latitudes, but not so greatly as in, say, butterflies or orchids. The list of 283 species (288 including five undescribed) known from Costa Rica (19,714 mi²) is 1.5x larger than the 192 species from the state of New York (54,555 mi²), while the Costa Rican butterfly list is 8x that of New York!

The situation is different in the case of damselflies, which have dramatically more species, genera, and families in the tropics than elsewhere. The great majority of odonates in the temperate zone are dragonflies of the suborder Anisoptera, whereas in the tropics damselflies (Zygoptera) are especially diverse. In Costa Rica, there are 9 damselfly families with 120 species and 5 dragonfly families with 170 species, including undescribed ones.

We still have much to learn about the biodiversity of odonates. New species continue to be discovered in Costa Rica and elsewhere in the tropics. Two new clubtail species were just discovered on the same stream in the Monteverde area of Costa Rica. Two swampdamsels (*Leptobasis*) described from Costa Rica in recent years were thought to be restricted to the country. One was subsequently found at several localities in southern Mexico and then in Florida; the other was discovered far to the north in western Mexico. They presumably occur in between but have yet to be found there. Two new species were added to the country list in one brief visit to Laguna del Lagarto near the border with Nicaragua in 2017, and an additional species was added in 2019.

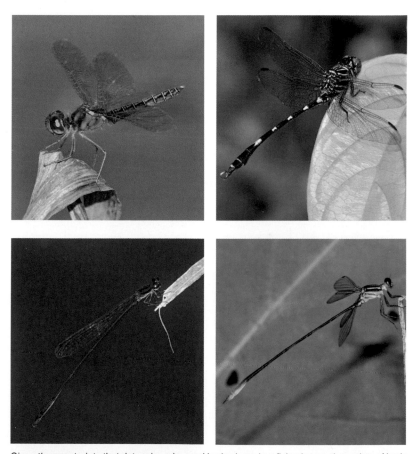

Given the constraints that determine wing and body shape in a flying insect, the variety of body forms in tropical odonates is impressive. Here are two dragonflies that show extremes in body form and coloration and two damselflies to show the different wing positions. Upper left Eastern Amberwing (*Perithemis tenera*); upper right Broad-striped Forceptail (*Aphylla angustifolia*); lower left Crimson Threadtail (*Protoneura amatoria*); lower right Mountain Flatwing (*Heteragrion majus*).

Rarity. Species in the tropics tend to occur in lower numbers than those in the temperate zones, and this is true for many organisms, not just odonates. Many species are known only from the original description of one or a few specimens from a single locality. While it is extremely unlikely that they are as limited as the few sightings suggest—otherwise they would have gone extinct long ago—it can be difficult to find some odonates even at the location where they were discovered. When a second locality is found, it may be far from the first one. Perhaps habitat preferences are so narrow in some widely distributed species that they thrive in relatively few places. After many years of fieldwork in Costa Rica, there are more than 20 species the authors have yet to see alive.

Seasonality. Those of us who live in temperate climates are used to seasonality, and it is a simple picture: cold in winter, hot in summer. Adult odonates fly about in summer but live in water as larvae or eggs during winter. As you move to lower latitudes, that picture changes. In southern Texas and southern Florida, for example, a surprising number of species can be found in midwinter. In tropical lowlands, it is warm all year, and there are some adult odonates always present. In Costa Rica, a majority of species can be found throughout much of the year, although some are distinctly seasonal.

Rainfall, not temperature, is the most significant variable when you're in the tropics. Indeed, Costa Rica has but two seasons, a wet and a dry season, though the timing and length of each varies throughout the country. In Guanacaste Province, the driest part of the country, several months can pass during the December to May dry season without a drop of rain. Water levels decrease, and many ponds and even streams dry up completely. In most years, regular rains begin in May, refreshing the landscape and filling the wetlands.

In wetter parts of the country, the southern Pacific and Caribbean slope, it can rain in any month, though some months are wetter than others and October is often the driest month of the year on the Caribbean slope (although the rainiest month of the year in the rest of the country). Even in the wetter regions of Costa Rica, odonates reach their greatest diversity and abundance in the rainy season. Many species appear as adults only in that season, and the difference in abundance of some common species may be an order of magnitude or more. This is especially pronounced in several stream-dwelling families, including clubtails, shadowdamsels, flatwings, and bannerwings, which may be entirely absent after November, then appear in abundance at the beginning of the rainy season in May. The basis of this pattern is not obvious, and biologists have no idea how their annual cycles are regulated. Perhaps emergence is stimulated by rising water levels, faster currents, or a decrease in water temperature after rain showers.

The driest region of Costa Rica is Guanacaste Province, in the northwest. The photo on the left is of a fry forest southwest of Cañas, taken on June 6th, at the end of the dry season; the second photo, of a nearby location, was taken August 8th, well into the wet season.

Migration. Although seasonal migrations are an important aspect of the natural history of many odonate species, few studies have been conducted on the subject. Migration involves the movement of individuals away from their place of emergence, usually to a different habitat and climate zone. Local displacement within a habitat, such as from one pond to another, is considered dispersal, not true migration. Dragonfly migrations are most often associated with habitats in which breeding sites dry up during part of the year. When ponds and streams dry out, the adults leave for another habitat that maintains permanent water; in some cases, they may fly to moist refuge sites at higher elevations or to humid river bottoms with evergreen forest, where they can wait for the wet season to arrive. In habitats with permanent water, many odonates pass the dry season as larvae.

Dragonfly migrations are spectacular affairs; hundreds or even thousands of individuals fly high in the sky or just skim the treetops, filing by in the same direction for hours or days. Within Costa Rica, the most common migratory path is between the Caribbean slope, where the rainforest has relatively little seasonal variation in weather, and the northern Pacific slope, characterized by dry deciduous forests that experience extreme seasonal variations in rainfall. The migrations take place in both directions. In the northwestern Pacific lowlands, when the first rains create temporary ponds and streams, dragonflies arrive to breed. At the end of the wet season, they migrate east, in noticeable numbers. In April and May, at the beginning of the rainy season, return migrations from east to west are rarely noticed, perhaps because numbers are low. Another eastward migration peak occurs in the middle of the wet season, during July and August, when there is a measurable dip in rainfall (referred to locally as the *veranillo*, or "little summer"). Also observed in *veranillo* is a second migration track that follows along the west coast, from the dry forests of the northwest to the wet forests of the south Pacific. The most conspicuous and predictable migrant groups include species of *Rhionaeschna*, *Gynacantha*, and *Triacanthagyna* among darners and *Pantala* and *Tramea* among skimmers. Seasonal migratory patterns for damselflies in Costa Rica have not been documented.

Elevation range contributes to diversity. As Costa Rica is a tropical country, its life zones are mostly a function of elevation and annual rainfall. Although vegetation types change dramatically with elevation, terrestrial habitats are not as significant as wetland habitats in the lives of these aquatic animals. The lower temperatures at higher elevations may be the greatest factor in determining changes in the composition of the odonate fauna, although the changes in aquatic habitats that occur at higher elevations may also play a role. With respect to odonates, the authors use the term *lowlands* to describe regions between sea level and 1000 m (3,280 ft); the terrain above that is called highlands, although the terms are not always precise. More than 40% of "lowland" species reach 1000 m or higher, in some cases considerably higher. Numerous lowland species have been found as high as 1450 m (4757 ft) at a pass in Monteverde, but perhaps only during migration. There are 22 species, less than a tenth of the fauna, that have so far not been found below 1000 m, and the authors consider these highland species.

Odonate habitats in Costa Rica. All but a very few odonate larvae live in water, and any body of water usually has odonates living in it. In tropical regions, many species are confined to forest streams or swamps (which are forested ponds), and it is these species that are most imperiled by logging and the conversion of forests to pastures and croplands. Some species are restricted to very special water habitats; streams vary in size, current speed, bottom composition, and forest cover, and these variables all apparently make a difference. About half of the Costa Rican species are stream species, indicating a great degree of specialization as well as the importance of flowing water.

Ponds and swamps also have their specialists, although these species tend to have wider distribution than stream species. Species of open areas also have broader ranges. Some of the common open-pond dragonflies of Costa Rica, for example, are common from the southern United States to Argentina and throughout the West Indies.

These photos show dragonfly habitats in Costa Rica, which include extensive marshes, large rivers, narrow streams, and tiny ponds. Even the bromeliads on tree branches furnish breeding sites for a few species of Odonata. a: Pond, Laguna del Lagarto; b: Marsh, Caño Negro; c: Forest pond, La Selva; d: Sarapiquí River, Selva Verde; e: Mountain stream, Tenorio Volcano; f: Slow lowland stream, La Selva; g: Swift lowland stream, Carara NP; h: Bromeliads, Villa Lapas.

Phytotelmata are leaf axils in bromeliads and other plants that are filled with water; this habitat is unique to the tropics. One helicopter damsel and one skimmer are confined to them in Costa Rica, and still other helicopter damsels breed only in water-holding tree holes. These odonates do not occur in wetlands, as one normally expects, but instead are scattered throughout the forest looking for their special water containers. Some of them stay in the canopy most of the time. A few darners (*Gynacantha*) and skimmers (*Libellula*) apparently use tree holes opportunistically but are not restricted to them.

Rainforest Odonates. Rainforests are hotbeds of odonate diversity. Many species breed in swamps, both permanent and seasonal, and forest streams, and they occur only within these dense forests. As most of these species do not occur in open country, cleared forests pose a challenge to conservationists. Also, many species that generally inhabit open wetlands retire to forests as immature adults during the dry season, presumably because prey populations are higher there, and the microclimate is cooler and more humid. They develop sexual maturity and mature coloration as the next rainy season begins, then leave the forest to breed.

Forest trails can be great places to find odonates, as they forage in the bright, warm treefall gaps and sunny spots scattered along them. Others are adapted to shade, even deep shade. Dusk-feeding darners are often flushed from their daytime roosts, and eventually one learns how to look for them before they fly off.

Odonate Biogeography in Costa Rica. Geologically, Central America was once isolated from both the South American and North American continents. Its biota thus had periods of isolation that promoted speciation. Some species came from the north, some from the south, and others evolved in situ. Of the 283 species discussed in this book, 32 are endemic to Costa Rica. Fifteen of these endemic species are shadowdamsels (*Palaemnema*) or knobtails (*Epigomphus*), both genera notorious for having species with restricted ranges. Some of the endemics likely also occur in southern Nicaragua or western Panama, but the fauna of both of those regions is less well known. An additional 15 species are known only from Costa Rica and Panama, most in the Talamanca-Chiriquí highlands, and six are known only from Nicaragua and Costa Rica. Not including these species with narrow distributions, 31 find their southern limit within Costa Rica, and 20 have their northern limit there. The country is clearly a mixing pot of odonates from north and south.

Because they occur in adjacent countries, any of the species listed in Appendix C with range boundaries in Nicaragua to the north or Panama to the south are worth watching for in Costa Rica.

Undescribed Species. At least seven undescribed species of Odonata occur in Costa Rica. They are in the following genera: *Palaemnema* (4), *Argia* (1), *Macrothemis* (1*), and *Perithemis* (1*). A few of them are fairly common but have yet to receive names. Species with an asterisk are included in the book. The world has too few taxonomists to handle the number of known insects needing description, not to mention those awaiting discovery!

Odonate Morphology

Body Structure. As typical insects, odonates have a head, thorax, and abdomen. The eyes take up a large part of the head, especially in dragonflies; a pair of tiny antennae and three light-sensitive ocelli are also situated on top. The "face" is made up of the *frons*, *postclypeus*, *anteclypeus*, and *labrum*, from top to bottom; the labrum covers a pair of *mandibles*, used to chew up prey. The jointed *labium* covers the mandibles from below. The thin neck membrane that connects the head to the prothorax allows maximum movement of the head, which can be rotated significantly more than 90°; this behavior can be observed when a dragonfly is watching for prey or cleaning its eyes with its legs.

Behind the head is a small segment that appears to be the "neck" but is in fact the first segment of the thorax, the *prothorax*, with the first pair of legs attached. The second and third parts of the thorax are fused into the *pterothorax*, which bears the four wings and the second and third pairs of legs. The pterothorax is enlarged because it holds the large wing muscles. The 10-segmented abdomen extends behind the thorax. Segment 2 of the male contains the secondary genitalia, the hamules with which he clasps the female abdomen, and the "penis" with which he transfers sperm. Segment 8 contains the *genital pore* of both sexes, segment 9 the *ovipositor* of those females that bear one, and segment 10 the terminal clasping appendages of males—the paired *cerci* (singular *cercus*) of all odonates and single *epiproct* (dragonflies) or paired *paraprocts* (damselflies).

Each leg is made up of a *femur*, *tibia*, and segmented *tarsus* ending in paired tarsal claws. The wings are thin, largely transparent, and heavily veined for support; the major veins, cells, and groups of cells are named and important in odonate classification. The small, thickened opaque area on the front margin of each wing near its tip is the *pterostigma* (which the authors call *stigma* for brevity), important in regulating the pitch of the wing during flight.

Dragonfly general morphology; male Ruddy Streamskimmer, *Elasmothemis cannacrioides*. S2-10 = abdominal segments (not all segments indicated on photo). S1 is very small, between the thorax and S2, and S3 is mostly hidden.

Dragonfly head and thorax; female Great Pondhawk, *Erythemis vesiculosa*.

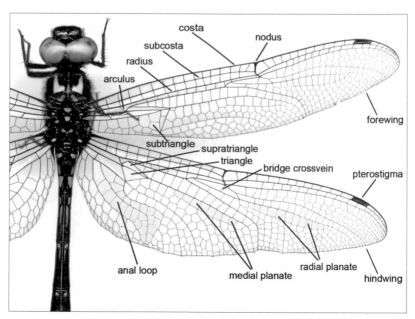

Dragonfly wings; male Masked Clubskimmer, *Brechmorhoga pertinax*. Major veins, cells, and groups of cells are labeled.

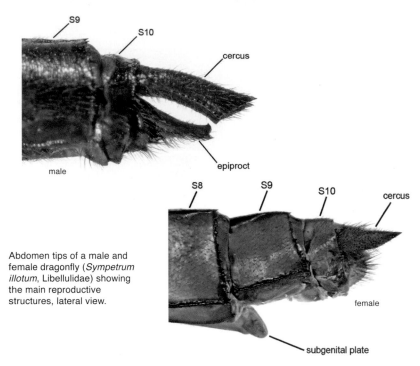

Abdomen tips of a male and female dragonfly (*Sympetrum illotum*, Libellulidae) showing the main reproductive structures, lateral view.

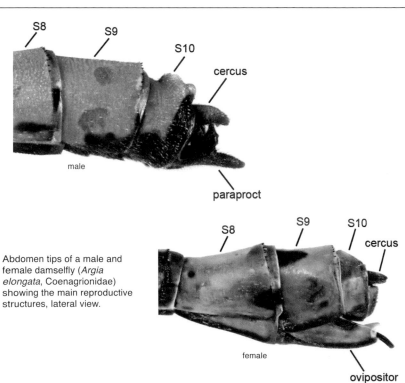

Abdomen tips of a male and female damselfly (*Argia elongata*, Coenagrionidae) showing the main reproductive structures, lateral view.

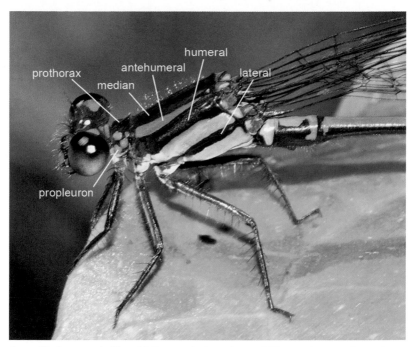

Damselfly thorax; male Waterfall Dancer, *Argia underwoodi*. The area between the prothorax and wings is the pterothorax, from which extend the wings and legs. The damselfly thorax is tilted backwards in comparison with the dragonfly thorax, so what is actually the front appears to be the top. The names on the pterothorax refer to the colored stripes often present on a damselfly thorax. S1–3 = abdominal segments.

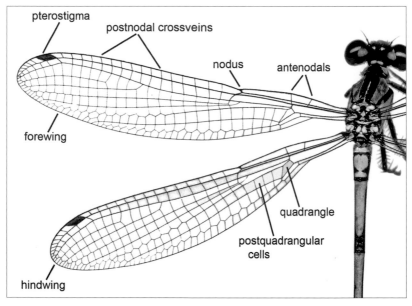

Damselfly wings; male River Dancer, *Argia insipida*.

Coloration. Odonates are among the most brightly colored insects. Males, which defend territories against other males and in some cases display to entice females to mate with them, are more brightly colored than females in the vast majority of species. Members of each family often share a general color pattern, although there are differences, often subtle, from species to species. Darners (Aeshnidae) show browns and greens that camouflage them when perched, with dashes of blue, especially in those that fly actively during the day. Clubtails are cryptically patterned, with stripes on the thorax and bands or spots on the abdomen. Many of the skimmers that fly up and down streams share clubtail patterns, as do some pond-dwelling skimmers such as speckled dashers and dryads. Other skimmers are the brightest of the Costa Rican dragonflies, with every color represented; some exhibit pruinosity, a gray waxy deposit that exudes from the cuticle to cover the surface and changes their appearance with maturity.

Among the damselflies, rubyspots show red and metallic coloration, pond damsels and shadowdamsels blue and black patterning, spreadwings metallic green plus pruinosity at maturity. Flamboyant and dusky flatwings show colors and behavior suggested by their names, and bannerwings are noteworthy for the iridescent wings of some species.

Polymorphism. Polymorphism, in which a population contains individuals with a different color pattern independent of sex and age, is well known in females of some odonate species. It occurs in a small percentage of Costa Rican species, including most dancers, bluets and forktails, the Black-tailed Darner, and the two largest dragonlet species, Black-winged and Band-winged. In these species a minority of females are colored and patterned more like males, while most have a less colorful, more camouflaged pattern. As striking exceptions, cascade damsels show wing polymorphism in both sexes, and multiple color morphs occur in males but not females in two bannerwing species.

Black-winged Dragonlet (*Erythrodiplax funerea*). Left, heteromorph female; right, andromorph (male-like) female.

How to Identify Odonata

The easiest way to learn to identify dragonflies is to capture them with an insect net and identify them in the hand, even using a hand lens to inspect the genitalia. However, to do this legally in Costa Rica you must obtain a research permit. The second-best way is to take good close-up photos of the individual in question, so that details not immediately evident in the field may be scrutinized at a later time in a magnified image. Always try to capture the entire color pattern, that is, from an angle that shows both the top and side. Good in-focus photos of the wings that show wing venation can help in identification. For some groups, clubtails and dusky flatwings, for example, getting a close-up photo of the terminal appendages is also helpful, indeed necessary in some cases.

Just as in birdwatching, the better look you get at a dragonfly, the more likely you will be able to identify it. The size and predominant color are, of course, the first things you will notice. Skimmers and pond damsels often can be identified on the basis of color pattern, sometimes quickly. But given that many species have similar coloration, you will often need more information. A first step, of course, is to identify the family. Is it a damselfly or dragonfly? If a damsel, are the wings closed or open? If a dragonfly, is it a flier or a percher? Check the eyes of dragonflies. If they are not touching, you've got a clubtail (if barely touching, a spiketail).

Identifying an odonate, like identifying a bird, often depends on subtle markings and the precise location of same. With dragonflies, for example, it makes a difference if that pale spot on the abdomen is on segment 7 or segment 8 or if a basal hindwing spot extends out as far as a particular vein or not.

Habitat and range can be very important clues to identification. Most species are adapted to live either in running or still water, so if you are on a stream, you can probably cut your potential species list in half. Some species are typical of open areas and are unlikely within a forest; others would only be found within a forest. And, because many species are so far only known from the Pacific or Caribbean slope of the country, location is important. Some 22 species occur only in the highlands, so if you are above 1000 meters in elevation, the list of possibilities is further restricted. The number of species declines with elevation, so above 2000 meters the list is even smaller. (Note that all measurements are in metric units only, to save space and to avoid ungainly numbers resulting from making conversions. Please see the Ruler and Measurement Conversion Chart on the inside back cover.)

It is important to understand the degree of variation that occurs in odonates. Even for a given species, individuals vary greatly in appearance from the time they emerge from the water until the day they die. The identification features described, unless otherwise indicated, refer to mature individuals of both sexes. Individuals that have not reached sexual maturity, called immatures, will almost always be differently colored, even if their markings are the same. Typically, but not always, immature males are colored like females. Finally, some species are sufficiently uncommon that even the authors have not observed all phases of their lives, with the result that descriptions of some of these species—especially females—may not be complete.

Odonate Photography

Not many years ago, the only way to learn to identify dragonflies and damselflies was to capture them, preserve the specimens, and key them out in technical keys. While this is still the most accurate method, especially in the tropics, regulations today in most countries require that one obtain research or collecting permits in order to collect specimens. The next best method is to substitute photographs for specimens, but this does not always guarantee a proper identification, as some characteristics are too small or concealed to be captured adequately in a photo.

As noted previously, anyone wishing to use photos to aid in identification should take them from all angles, and at close a range as possible to register fine details. Terminal appendages are important in males, and both dorsal and lateral shots may be needed. The very pleasing dorsolateral photos of dragonflies typically taken may have the thoracic pattern obscured by the wings, so effort should be made to show that often critically important pattern from the side. Finally, wing venation can be quite useful as well, so good photos of the wings perpendicular to their surface are also just as important as that classical three-quarter view.

Dragonflies are usually wary, so using a telephoto lens is the best way to assure good photos. Damselflies are often tamer, but even they don't usually tolerate a very close approach. A slow approach is always best, even when at some distance. Damselflies with closed wings are most easily photographed from the side, as even a narrow depth of field will capture the whole animal in focus. Dragonflies and damselflies with their wings open will have parts out of focus unless the right technique is used. Closing the aperture is the best way to increase the depth of field, and some newer cameras have in-camera focus stacking that will do the same.

Collecting and Collections

Substantial collections of Costa Rican odonates exist, providing the basis for this book and future study. At this time general collecting is not permitted in the country, and anyone wishing to conduct specimen-based research or surveys must apply for a research permit from the Ministerio de Ambiente y Energía. The relevant website is

http://www.sinac.go.cr/ES/tramitesconsultas/permisoinvestigacion/Paginas/default.aspx.

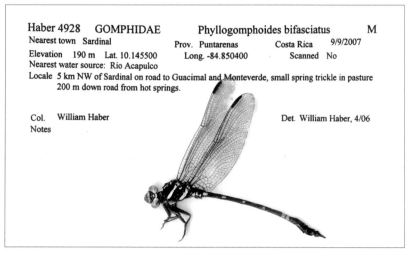

Haber 4928   GOMPHIDAE   Phyllogomphoides bifasciatus   M
Nearest town   Sardinal   Prov.   Puntarenas   Costa Rica   9/9/2007
Elevation   190 m   Lat. 10.145500   Long. -84.850400   Scanned   No
Nearest water source: Rio Acapulco
Locale   5 km NW of Sardinal on road to Guacimal and Monteverde, small spring trickle in pasture 200 m down road from hot springs.

Col.   William Haber   Det. William Haber, 4/06
Notes

Well-preserved specimens with accurate collecting data are the basic elements of our knowledge of Costa Rican dragonflies and damselflies.

Conservation

Dragonflies and damselflies are doing well relative to many other insect groups, perhaps for two reasons. First, they are generalized predators and so are not dependent on particular plant or animal species, which makes them better able to persist in a changing environment. Second, they live at and near water and are thus shielded somewhat from some climate-change problems such as extreme heat or fire. Nevertheless, conservation is as critically important for dragonflies and damselflies as it is for other organisms. These insects depend on adequate aquatic habitats for their larval lives and adequate terrestrial habitats for their adult lives. Drought is a serious threat for those species that live in shallow wetlands or phytotelmata, just as heavy rains pose a problem for species that live in running water, as they can wash out streams and rivers, eliminating entire larval communities.

Costa Rica has an exemplary number of reserves that conserve natural habitats and their biota, particularly important given the heavy use of agricultural pesticides in the country. And people continue to clear land and destroy habitats in other ways, for all the usual reasons. Everywhere, maintaining clean, unpolluted freshwater systems and conserving forests is the prescription for saving our odonates. Conservation efforts need to take into account both larval and terrestrial habitats and their proximity to one another.

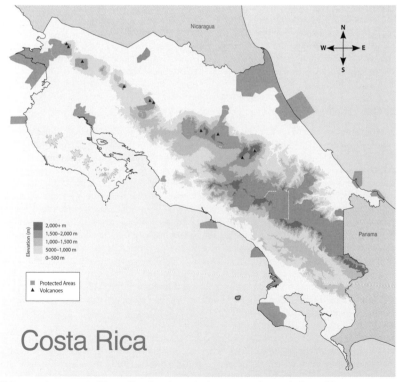

This map shows most of the national parks and other protected areas in Costa Rica.

Using the Book

Family and Genus Accounts. In addition to species accounts, this book describes every odonate family that occurs in Costa Rica. When a genus contains more than one species, we also include a description of that genus. Reading family and genus accounts will impart familiarity with general aspects of each group, including habitat, behavior, and appearance. In some family accounts, you will find considerable information that will help identify the species in that family.

Order of Species. In books on Odonata, it has been the custom to list families in some semblance of evolutionary order, from earlier to later evolved. Listing of species has been more variable. Some checklists list odonates in alphabetical order by genus and species within each family, eliminating the need to think about evolutionary order (*phylogeny*). In this book, the authors attempt to list species in some semblance of phylogeny, acknowledging that current knowledge of this is improving with the addition of recent genetic analysis but still far from complete. The authors follow the order established in some earlier books and also presented for North American species in the pair of field guides by Paulson (listed under Bibliography). The tropical genera and species not included in those books are interpolated as neatly as possible.

The order of presentation does not agree entirely with some of the latest publications on molecular phylogeny of odonates. Gliders (*Pantala*) and saddlebags (*Tramea*), for example, have long been considered close relatives, but genetic analysis indicates they are not so closely related. They have been kept together in this book, partly as a nod to tradition and partly because they are similar in appearance.

Appendix B contains a list of all Costa Rican species, including the author of the species and year of description. For convenience, families are listed in phylogenetic order and genera and species in alphabetical order (thus different from the arrangement in the book).

Species Accounts

Photographs. In the spirit of "a picture is worth a thousand words," we have chosen to rely often on photographs to indicate key morphological features. Gathering photos for the illustrations of a given species was perhaps the most difficult task in producing this book. The authors examined thousands of photos and tried to choose the best of the lot. When live photos were not available, scans were selected from those that Haber has been accumulating for years in Costa Rica. In a fairly small number of cases, when neither a photo nor a scan was available, the authors have used a scan of a specimen, which is not ideal, as specimens do not show the coloration of live individuals accurately. In addition, a few photos are of recently emerged individuals (tenerals), which show the pattern but not the color of the adult. Photos of specimens and tenerals are indicated as such. There remain 14 Costa Rican species for which there are no photos of live individuals, as well as other species in which only one sex has been photographed. The authors hope this book will stimulate not only odonate enthusiasts but also nature photographers to find and photograph these little-known species.

Please note that not all photographs on a two-page spread are to scale, sometimes not even on the same page.

Names. For all the odonate species in this book, the authors indicate common (English) and scientific names. While scientific names have been used by odonatologists for centuries, common names were coined relatively recently—in 1996 for species in the US and Canada and only in the last few years for species south of the US-Mexican border. Note that the accent marks have been dropped from common names of Spanish origin: Chirripo, Limon, and Reventazon. Professionals have always used the scientific names, as that leads to accurate communication no matter the language of the speaker or writer, while the English common

The above, left to right, are Roseate Skimmer, Carmine Skimmer, and Gray-waisted Skimmer.

names were created to be used by naturalists for whom odonates might not be a primary interest. Many people who look at dragonflies prefer the common names, and we hope to facilitate odonate watching for all who spend time in nature.

Scientific names are a handy invention indeed, allowing the naming of all organisms in a universal language (most of the names stem from Greek and Latin, of course). The first of the two names is the *genus* (plural, *genera*), and all members of that genus share that name. Thus, we can easily see that *Orthemis ferruginea* and *Orthemis discolor* are more closely related than either of them is to *Cannaphila insularis*. Their common names Roseate Skimmer, Carmine Skimmer, and Gray-waisted Skimmer, respectively, do not offer that information. In addition, with so many similar species in some groups, coming up with unique common names has proven difficult, so a given common name does not always show a unique characteristic. Note that in this book common names of species are capitalized.

All odonate genera in this book have also been given common names, which are not the equivalent of proper names and are thus written in lower case. For example, *Orthemis* are called "tropical king skimmers." Some of these names are sufficiently lengthy that when one flies past, it is easier to say "there goes an *Orthemis*" than "there goes a tropical king skimmer." Acknowledging that, the authors sometimes use the real generic name in the text when discussing identification or behavior as part of an effort to help people become more familiar with them. In addition, and this may be important to some, the primary literature on Odonata taxonomy, distribution, ecology, and behavior uses only scientific names.

Measurements. For each species, the text lists both the average total body length (TL) and hindwing length (HW) in millimeters only to save space (please see the Ruler and Measurement Conversion Chart on the inside back cover). These measurements refer to males, not out of any ingrained sexist impulses but because males are much more commonly seen than females. All species vary in size, but generally no more than 5–10%; thus, a total length of 50 mm indicates a potential range of about 45–55 mm. Note that females have a relatively shorter abdomen than males and are a bit heavier-bodied, which results in a slightly greater hindwing length (the wings support the weight, after all). In some cases, the calculated "average" measurements were based on only a few individuals; thus, a difference of a few millimeters between two related species could be more an artifact of an erroneous average than a real difference.

One interesting thing to note about odonate sizes in the tropics is our observation of a slight increase in size with elevation; the authors have documented this in relatively few species, so more information is needed to make conclusive statements. On a related note, within a given group, the montane species are often the larger ones, notably so in flamboyant flatwings, rubyspots, and dancers.

Identification. The species accounts begin by pointing out the most important features for identifying a species, followed by ways to distinguish it from similar species. In some cases, this is brief, because identification is relatively straightforward. Accounts may be lengthier when a species can be confused with several species or is very similar to one or more of them, with additional details being necessary for positive identification. Descriptions, species comparisons, and statements such as "no other species has..." refer only to the situation in Costa Rica.

It is always good to identify the sex of an individual, as so many males and females within the same species look different. Males tend to be slender, but the accessory genitalia in segment 2 of the abdomen create a bulge, visible from the side and often quite prominent, that is apparent in almost all species. Females lack this. The tip of the abdomen is usually more pointed in males, and their distinctive appendages can be perceived: a pair of cerci in all and below them the epiproct of male dragonflies or paraprocts of male damselflies. Females are generally heavier-bodied than males, as they have to have space for large numbers of eggs; they are also slightly shorter. Female damselflies and darners all have ovipositors, which are usually obvious. Female clubtails, spiketails, emeralds, and skimmers lack an ovipositor, and their abdomen typically looks blunt-ended in comparison with the male of the same species. They have a pair of rather short cerci at the end, apparent on close inspection.

The descriptions generally apply to mature individuals. Immatures differ from reproductive adults in color patterns, a little or a lot, more so in males than females. Immature and mature female colors tend to be muted and serve as camouflage, while in mature males the brighter colors function in display. The bright colors of adults are very often duller and paler in immatures and go through a gradual change with age. The patterns of spots and stripes in general change less than the colors.

For some genera, the authors include illustrations or photographs of male appendages to aid in identification. Note that the species accounts are the only sections written in somewhat telegraphic style. Costa Rica is abbreviated as CR and the United States as US.

Habitat and Behavior. A description is given for the general habitat for each species, which is to say its breeding habitat; as you will read, some species prefer still water, others flowing water, in either open country or forest. Perching sites are also mentioned. The authors either describe the behavior of a specific species or, in some cases, describe the general behavior of all members of a genus. Natural history notes are included for some species, for interest and also as a possible additional aid in identification.

Range. The authors describe for each species both the range within Costa Rica and the entire range, including information about elevation. For common species, there is abundant range information, but this is not the case for rare species. When only a single elevation is given, this indicates that there is insufficient information to give an elevational range.

Species Accounts

Twigtails (family Perilestidae)

Two species. Members of this neotropical family all look quite similar, with wings held partially open and an exceptionally long, vaguely ringed abdomen; note upcurved tip of abdomen hanging down over some small forest stream or in the forest understory, looking like a vine or twig. The female abdomen is shorter, with a bulbous end to accommodate the ovipositor. Twigtails occur in both Pacific and Caribbean lowlands, but both species are poorly known and uncommon. Little is known of their behavior. When away from the water, they often perch on vertical stems within 30 cm of the ground, especially in sites with vine tangles or dense shrubs and herbs around them, either in the understory or in light gaps as far as 50 m from a stream. When disturbed, they fly straight up to become lost within the vines. Males perch up to a meter above the water on leaves or twigs overhanging quiet pools in forested streams, including pools below small waterfalls.

Horned Twigtail *Perissolestes magdalenae*

TL 55 mm; HW 23 mm. **Identification:** Males of the two twigtails look essentially the same, with alternating whitish, black, and orangish bands on middle abdominal segments. In this species, pale markings on thorax very pale blue; in Green-striped, if color is visible, it is tinged with green. Green-striped has fewer pale bands on abdomen, those on S7-8 absent or at least less obvious than in Horned. Male cerci should be seen to confirm identification. In Horned, each cercus has an inner tooth at about half the length of the cercus; note slight indentation toward tip of tooth. On Green-striped, inner tooth is located closer to base of cercus, with a more pronounced indentation. Female Horned with conical projection on posterior lobe of prothorax, and ovipositor valves do not extend beyond tips of cerci. Female Green-striped lacks prothoracic horn and has slightly longer ovipositor than Horned, clearly extending beyond tips of cerci. **Habitat and Behavior:** Swampy spring seeps and small streams to rocky rivers in forest. **Range:** Lowlands of both slopes, to 900 m. Eastern Mexico to Colombia.

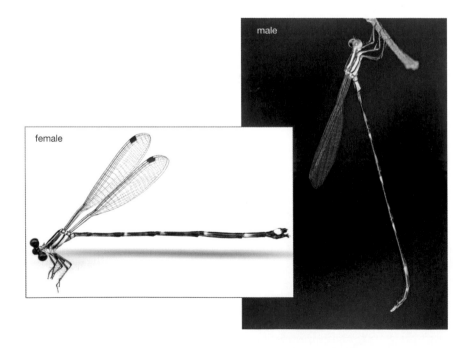

male

female

TL 53 mm; HW 23 mm. **Identification:** Nothing else is even vaguely similar to these elongate ringed damselflies. See Horned Twigtail. **Habitat and Behavior:** Small streams in forest. **Range:** Lowlands of Caribbean and south Pacific slopes, to 600 m; so far known from only scattered locations. Nicaragua to Peru.

male

female

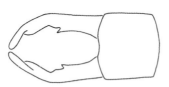

genus *Perissoletes*
male appendages in dorsal view

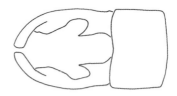

Horned Twigtail	**Green-striped Twigtail**
Perissolestes magdalenae	*Perissolestes remotus*

Spreadwings (family Lestidae)

Nine species. Spreadwings are large to very large damselflies, most with an entirely dark head except for blue eyes and blue labrum; mature males of many species have a pruinose abdomen tip. Spreadwings perch with wings held open partway and abdomen hanging below the horizontal. The stigmas are relatively longer than in pond damsels. Males have forcipate (forceps-shaped) cerci, longer than those of other large damselflies that hold their wings open. They can be distinguished from other open-winged damselflies, especially the somewhat similar dusky flatwings (*Philogenia*), by their blue eyes when mature.

STREAM SPREADWINGS   genus *Archilestes*

Three species. The three stream spreadwings are large; even the smallest of them (Great Spreadwing) is reminiscent of a dragonfly when flying over water. All have bright blue eyes, duller blue or even brown in females. Only Great is common and widespread; the other two are forest-stream dwellers of mid- to high regions.

Great Spreadwing   *Archilestes grandis*

TL 59 mm; HW 32 mm. **Identification:** Much more abundant and widespread than the two other stream spreadwings, and also significantly smaller (though still much larger than any pond spreadwing and most other CR damselflies). Neither male Yellow-eyed nor Cloudforest (p. 28) has a pale abdomen tip; in both sexes of both species, the thorax is more noticeably metallic green. **Habitat and Behavior:** Common at small to medium upland streams, often in the open, unlike its close relatives. Males mostly at pools, even isolated ones, perching over and at times well above the water. Females more often oviposit in woody stems of shrubs and trees but also in the stems of herbs and grasses, up to a considerable height above water. **Range:** Both slopes, to 1450 m; more common in highlands. Occurs across US and south to Venezuela.

male

female

TL 66 mm; HW 34 mm. **Identification:** Distinctly larger than Great Spreadwing, with lower surface of eyes yellow and no pruinosity; thorax metallic green and chartreuse. Paraprocts not evident in males, small but visible in males of the other two species. Lower sides of thorax yellow in female Yellow-eyed; dark with two yellow stripes in female Great and greenish with dark stripe in female Cloudforest (p. 28). **Habitat and Behavior:** Males perch below canopy, 1–3 meters above pools in small streams and above isolated pools with sun exposure, in or near forest. **Range:** Both slopes, 400–900 m. Eastern Mexico to Costa Rica.

male

female

TL 67 mm; HW 38 mm. **Identification:** One of the largest damselflies in CR. Distinguished from other stream spreadwings by vivid emerald and blue-green stripes on thorax; note dark stripe low on each side. Male cerci slightly curved downward and with broad tips. Great Cascade Damsels (p. 78), with black and yellow thorax, hold their very large wings closed. **Habitat and Behavior:** Small streams in forest. Quite unusual, in that females are seen much more often than males, primarily when they come to the water to oviposit. Mating may take place far from the breeding habitat, perhaps in the canopy. Females sometimes territorial at quiet pools; they oviposit solo (unusual for a spreadwing), in the stems of a variety of plant species, either low (herbs) or to 2 m high (in trees), and usually slightly back from the water's edge. **Range:** Cloud forest in northern mountain ranges; also in Braulio Carrillo NP; 700–1600 m. **Costa Rican endemic.**

female

---

genus *Archilestes* (p. 26)
male appendages in dorsal view

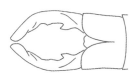 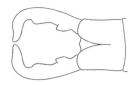 

Great Spreadwing	**Yellow-eyed Spreadwing**	**Cloudforest Spreadwing**
Archilestes grandis (p. 26)	*Archilestes latialatus* (p. 27)	*Archilestes neblina*

POND SPREADWINGS   genus *Lestes*

Six species. Pond spreadwings are locally common at ponds and marshes, occurring in lowlands and up to highlands. At least some of them spend the dry season in woodlands, near sites where seasonal ponds form in the rainy season. All are similar in size, and all males have blue eyes, a black head with blue labrum, a dark abdomen tipped with gray, and forcipate cerci. Females usually have eyes paler blue and tan underneath, lack pruinosity (female Chalky and Tikal become pruinose with age), and have an expanded abdomen tip. Color patterns vary with age, and immatures with brown eyes are commonly seen away from breeding sites, showing patterns that may be obscured with maturity. Male appendages may be important for identification. Four of the six species also occur in the US.

Rainpool Spreadwing   *Lestes forficula*

TL 38 mm; HW 19 mm. **Identification:** Thorax pale blue or greenish (brown in immature); two pairs of narrow metallic green stripes are distinctive for both sexes and all ages. In this, Chalky (p. 31), and Tikal (p. 30), male paraprocts almost as long as cerci (shorter in the other three species). Chalky has distinctive outcurved paraprocts; those on Rainpool and Tikal are straight or slightly incurved and quite similar. In Rainpool, tips are distinctly expanded in dorsal view, in Tikal somewhat less. Cerci also differ; in Rainpool, note curve between the two inner teeth (in Tikal, fairly straight there). **Habitat and Behavior:** Marshy ponds in open areas. Males perch at edge of vegetation rather than within it; they fly over open water, often among large sedges. Pairs oviposit in tandem. Like most spreadwings, adults become more active later in the day, when pairs also are more evident. **Range:** To date, known only from northern Pacific lowlands and Caribbean slope of the Guanacaste Mountain Range, to 600 m. Likely elsewhere, at least in Caribbean lowlands. Gulf Coast of US south to Peru and Brazil; also occurs in West Indies.

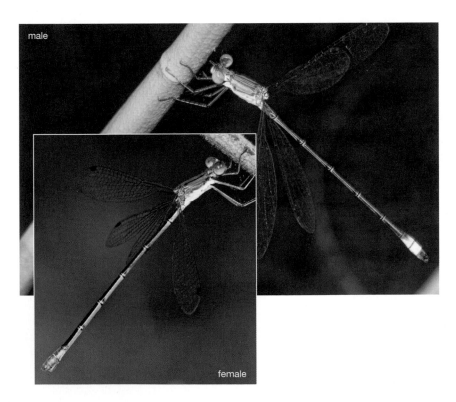

male

female

TL 39 mm; HW 20 mm. **Identification:** Pruinose thorax distinguishes both sexes from Blue-striped; they often occur together. Tikal rarely overlaps with Chalky, the other pruinose species; on Tikal note faint metallic median and humeral stripes lacking pruinosity (entire thorax pruinose in Chalky). Male Chalky has S2 half pruinose, not at all in Tikal. Also note outcurved paraprocts in Chalky, straight in Tikal. Pruinosity may cover thorax and abdomen tip in female Tikal, but usually to a lesser degree than in Chalky. **Habitat and Behavior:** Natural and artificial ponds in forest, usually open and sunny; found in both seasonally dry and permanent ponds. Males often perch conspicuously on emergent stems but are not strongly territorial. Females oviposit in tandem into live plant stems 10–30 cm above water. Both sexes sometimes found in herbaceous vegetation well away from water. **Range:** Both slopes, to 1400 m. Southern Mexico to Panama.

TL 41 mm; HW 21 mm. **Identification:** This is the most pruinose spreadwing; when mature, pruinosity covers entire thorax and rear of head of both sexes; pruinosity more extensive at abdomen tip than in other species. Male cerci longer than in other spreadwings and curved downward near tips; paraprocts two-thirds length of cerci, diverging at tip, so somewhat S-shaped from above. Less pruinose females can show hint of pattern on top of thorax. Immatures distinctive, with dark spots on plain brown thorax and bicolored brown and tan stigmas. **Habitat and Behavior:** Found in open areas in ponds and temporary rain pools. **Range:** Lowlands of Northern Pacific slope, to 300 m. Southwestern US to Costa Rica.

TL 40 mm; HW 21 mm. **Identification:** Mature individuals identified by blue or tan antehumeral stripe that is slightly convex on outer edge and narrows to point above. Paraprocts two-thirds length of cerci. Sides of thorax vary from tan in immatures to dark brown to blackish in mature males, becoming partially covered by pruinosity with age. Only spreadwing in which thorax can look entirely pale (except for black median stripe). **Habitat and Behavior:** So far, known from few localities in Costa Rica. Occurs at open mountain ponds, males and pairs in emergent sedges and grasses above open water. **Range:** Central Valley and Central Mountain Range, 1400–1900 m. Southwestern US to Costa Rica.

male

female

TL 43 mm; HW 22 mm. **Identification:** Only lowland spreadwing with noticeably black and blue striped thorax in both sexes; only S9 pruinose in males (pruinosity more extensive in males of other species). Paraprocts half length of cerci. Dark thoracic colors metallic green to bronze. Antehumeral stripes with same form as in Plateau Spreadwing, but humeral stripe is more distinct and never obscured by pruinosity. **Habitat and Behavior:** Permanent and temporary ponds, both natural and artificial, with much grass or sedges; found in the open and in forest (usually in sun in forested areas). Males maintain perches on emergent stems, often in dense tall grass. Females oviposit in tandem in live plant stems, 10–30 cm above the water. Immatures scattered throughout wooded habitats. **Range:** Found on both slopes, to 1400 m. Texas, Florida, and West Indies south to Ecuador and Suriname.

male

female

TL 48 mm, HW 27 mm. **Identification:** Largest of pond spreadwings. Like large upland version of Blue-striped Spreadwing (p. 33), but males with broader antehumeral stripes and pruinosity on S9-10 (only S9 in Blue-striped). Dark thoracic stripes black (metallic green in Blue-striped, easily visible in sun). Females also distinguished by more prominently striped thorax. In both sexes, rear of head black (pale in Blue-striped). **Habitat and Behavior:** Males perch in sun, guarding temporary and permanent ponds and marshes in cloud forest. Also found at quiet pools in streams and at pools formed in root holes where large trees have fallen. Males interact with still larger male Great Spreadwings, both occurring where marshes and streams meet. **Range:** Occurs in cloud forest, 1200–1700 m. **Costa Rican endemic.**

female

male

genus *Lestes* (p. 29)
male appendages in medio-dorsal view

Plateau Spreadwing
Lestes alacer (p. 32)

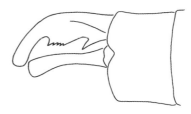

Rainpool Spreadwing
Lestes forficula (p. 29)

Montane Spreadwing
Lestes henshawi

Chalky Spreadwing
Lestes sigma (p. 31)

Blue-striped Spreadwing
Lestes tenuatus (p. 33)

Tikal Spreadwing
Lestes tikalus (p. 30)

Shadowdamsels (family Platystictidae)

Eleven species. All are in a single genus, *Palaemnema*, with 11 described and at least four undescribed species in CR. The undescribed species, scattered around the country, make it even more difficult to make an accurate identification, which ultimately may depend on careful examination of male appendages (see illustrations). Geography can play a part, as some species are known from only one side of the country. They do not seem to prefer distinct habitats, although note that Nathalia can be found at the largest and most sunlit streams frequented by shadowdamsels.

These damselflies of forest streams are larger than most pond damsels. Both sexes can be recognized as shadowdamsels by clear wings held closed well above their slender, pale-ringed abdomen and very long tibial spurs, even longer than in dancers. Wings may be held partly open when individuals are disturbed. Males have complex, species-specific appendages; females have a bulging abdomen tip that often has a pair of pale spots on S9. On males, note a black thorax marked with either green or blue; five species have antehumeral stripes, and all but three have a blue abdomen tip (S8-9). Tenerals with yellowish thoracic colors are particularly difficult to name, and the colors can change to green and then blue with age. Only Cacao and Reventazon are reliably green when mature. Ink-tipped is the sole species with black wing tips. In lateral view, propleuron (lateral part of prothorax immediately above fore coxa) color in both sexes can also be used to divide the group—in four (listed first) it is pale, in seven black. Females can be quite difficult to distinguish with the present state of knowledge, but certain species have distinctive characteristics, such as stigma shape, presence or absence of antehumeral stripe, pale spots on abdomen tip, ovipositor length, and eye color patterns.

As many as nine species have been found at the same site, but in most sites only one to three are present. They perch in the shaded understory along forest streams, from lowlands well into the mountains, at least to 1700 m. They are often in dense vegetation at tiny trickles, where they are difficult to photograph and can be collected only by hand.

A variety of reproductive behaviors has been observed, although only two species are well studied. Some species use a lek mating system in which males gather together among shrubs over a stream to await females. Others hold solitary territories along streams. Many have not been observed at water at all and may arrive only at certain times of day. After mating, pairs may stay together for hours, with males guarding egg-laying females by perching next to them. The larvae, which live on and under rocks and stones in gravel riffles and sometimes in fast-flowing water, are pale colored and look rather like termites.

TL 43 mm; HW 23 mm. **Identification:** Propleuron pale; lacks antehumeral stripes. This species has a blue abdomen tip and is one of five with thorax entirely black above in males. It differs from all other shadowdamsels in having short wide stigmas in both sexes, only slightly longer than wide. Females have pair of small blue spots at mid-length on upper side of S9 (basal in all others). Also unique in having exceptionally long ovipositor, extending perceptibly beyond abdomen tip. **Habitat and Behavior:** Small streams in forest. Males of this species observed perching on dead twigs about 1 m above water at small, rocky streams. **Range:** Caribbean slope, so far known only at 800–950 m. **Costa Rican endemic.**

male

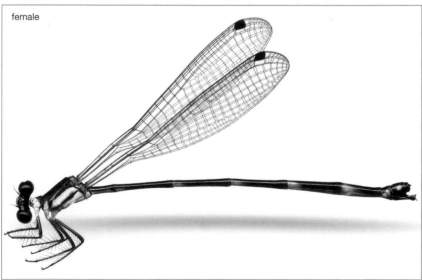

female

TL 44 mm; HW 23 mm. **Identification:** Propleuron pale; lacks antehumeral stripes. Thorax that is entirely black above and blue abdomen tip make males look like Chiriquita (p. 37), but Arch-tipped has long, narrow stigmas, clearly two times as long as wide. On females, pale spots on each side of base of S9 are inconspicuous to virtually absent. **Habitat and Behavior:** Small forest streams. While other Costa Rican species in this group typically perch low in more open understory, this species is often found among clusters of large-leaved herbs. Males perch 5–10 m away from water, females up to 30 m. Neither sex seen at water to date. **Range:** Both slopes, up to 900 m. Costa Rica and Panama, but differs from Panama population and may be a distinct species.

male

female

Black-backed Shadowdamsel   *Palaemnema melanota*

TL 48 mm; HW 25 mm. **Identification:** Propleuron pale. This species, known only from the type specimen, is distinguished by its appendage structure. One of five species lacking antehumeral stripes; it may have a dark abdomen tip, but that was not clear in the poorly preserved specimen. Must be rediscovered. **Habitat and Behavior:** Presumed to live along forest streams. **Range** Caribbean slope; known only from Tuis (1000 m), east of Turrialba. **Costa Rican endemic.** Not illustrated.

Nathalia Shadowdamsel   *Palaemnema nathalia*

TL 42 mm; HW 25 mm. **Identification:** Propleuron pale. The showiest member of its genus, with extensive bright blue on thorax and with wide antehumeral stripe. Dentate (p. 41) and Carrillo (p. 42) similarly patterned but with less blue on thorax and black propleuron. Wing tips smoky but never black as in Ink-tipped. Females have greatly reduced humeral stripe and thus look like females of Dentate and other shadowdamsels with unstriped thorax; paired markings on S9, as in most other species. **Habitat and Behavior:** Forest streams and rivers. The most abundant and widespread shadowdamsel on the entire Pacific slope; typically most common along medium-sized streams and small rivers (3–15 m wide) with fast currents and rocky bottoms. Quite rare on Caribbean slope. Only shadowdamsel that can be found outside of forests, among tree buttresses in pastures, for example. **Range** Pacific slope, to 1200 m; Caribbean slope, to 300 m. Southern Mexico to Venezuela.

male

female

39

TL 47 mm; HW 27 mm. **Identification:** Propleuron black. Males distinctive in having green thorax, very narrow antehumeral stripes, and abdomen tip dark except for yellow or bluish basal spot on either side of S8. Females have narrow antehumeral stripes and pair of pale spots on S9 that may meet above, with posterior edges more jagged than in other species. **Habitat and Behavior:** Small streams and spring seeps in forest. During the morning, territorial males perch at stream margins on stems or twigs 10 cm to 1 m above the water, where face-to-face fights often occur. Females and immatures perch near forest floor 5–30 m away from water, often next to small light gaps. **Range:** Both slopes, 800–1700 m. **Costa Rican endemic.**

TL 42 mm; HW 24 mm. **Identification:** Propleuron mostly black. One of six shadowdamsels with distinct antehumeral stripe in male. Males distinguished from Ink-tipped (p. 45) by clear wings; from Nathalia (p. 39) by thorax that is generally greenish, not blue (although some Dentate individuals have blue thorax); dark propleuron; and smaller size; and from Cacao by abdomen tip with extensive blue. Males easily confused with Carrillo (p. 42), which prefers smaller streams, but the latter has all-black postclypeus and dorsal tooth on cercus visible from side; must check appendages to be certain. Female stigmas and ovipositor shorter than in Carrillo; also, antehumeral stripe reduced to basal dash or interrupted in middle (usually complete in Carrillo), and S9 is less bulging. **Habitat and Behavior:** Found at medium-sized rocky streams and rivers with swift flow in primary and secondary forest; often locally common. Also emerges from spring trickles down steep rocky slopes, so may be highly adaptable. Dentate can overlap with Carrillo where streams of different sizes come together, but Dentate is usually more abundant. **Range:** Both slopes, to 1200 m. Costa Rica and Panama.

male

female

41

TL 45 mm; HW 27 mm. **Identification:** Black propleuron and antehumeral stripes on green to blue thorax make this look exactly like Dentate (p. 41). Abdomen tip blue; similar to Ink-tipped (p. 45), but wings clear. Females with mostly complete antehumeral stripe; basal cream to blue dorsolateral spots on either side of S9. **Habitat and Behavior:** Small slow streams in forest. Males perch on stream margins but also form leks in shrubs overhanging water. Immatures in understory and light gaps away from streams. **Range:** Both slopes, to 900 m. Nicaragua and Costa Rica. First described from a population in Braulio Carrillo NP on Caribbean slope. Other populations found at lower elevations throughout the country. (The description refers to Braulio Carrillo population. Other populations with similar appendages have postclypeus ranging from partly blue to all blue and S8-9 that varies from partly blue-spotted to entirely black. These different populations may be incipient or distinct species that require further study to delineate genetic relationships.)

male

female

TL 60 mm; HW 28 mm. **Identification:** Propleuron black, antehumeral stripes absent. Longest shadowdamsel in CR; males with very long abdomen, about 1.8x hindwing length (on other species, 1.5–1.6x hindwing length). Thorax entirely black above; blue on sides with black stripe below that runs to base of hindwing (on some other species that area is blue); pale tan below stripe. Females longer than females of other species; thoracic pattern similar to that of males. Unique among females in having S8-10 sky blue above; when young, only S9 with faint pale dorsolateral spot. **Habitat and Behavior:** Very small streams and even swampy seeps in primary or secondary forest. Prefers very small, slow spring trickles less than 1 m wide, where there are few rocks. Mature males may be found within a meter or two of stream margins. Immatures and females usually perch low in sunny spots in forest understory, 5–20 m from stream. **Range:** Caribbean slope, 300–800 m. Nicaragua and Costa Rica.

male

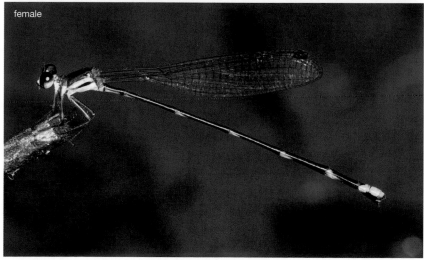

female

43

TL 40 mm; HW 22 mm. **Identification:** Propleuron black. Another shadowdamsel that lacks ante-humeral stripes. Male abdomen has blue stripe low on side of S8, large dorsolateral blue patch on S9, and small blue spot on each side of S10; however, extent of blue on abdomen variable with age, with immature males starting out black. Could be confused with Arch-tipped (p. 38) and Chiriquita (p. 37), both with S8-9 blue above, but latter two not known to overlap geographically with Janet's. Females lack an antehumeral stripe, show a pale spot on side of S9, and ovipositor extends slightly beyond cerci. Note that an undescribed species similar to Arch-tipped with pale propleuron occurs in same range as Janet's. **Habitat and Behavior:** Small slow streams in forest. **Range:** Known only from Osa Peninsula and adjacent southern Pacific region. Costa Rica and Panama.

male

female

TL 47 mm; HW 27 mm. **Identification:** Propleuron black. Bold antehumeral stripes in both sexes. Black wing tips of mature males make this the most easily identified shadowdamsel in CR. Immature males lack black tips, but they and females have short dusky to black stripe low on rear of thorax (apparently the only species to have this). **Habitat and Behavior:** Medium to large streams in forest. Males often perch near water at base of trees with large buttresses. **Range:** Both slopes, to 700 m. Nicaragua and Costa Rica.

male

female

TL 50 mm; HW 28 mm. **Identification:** Propleuron black. Distinguished by being very dark, including prothorax and abdomen. Thorax entirely black above but sometimes with almost imperceptible antehumeral stripe reduced to spot at base. Lower sides dull tan or cream to green. A bit larger than most other species. Pale orangish triangles on abdomen located low on side (not visible from above). Female thorax patterned like that of male; also note dull yellow spot on lower half of S9 and ovipositor shorter than cerci. **Habitat and Behavior:** Very small streams and springs in forest. A male was observed in primary forest, perched among dead twigs of a fallen tree in a small gap near a spring seep. Another male was perched near water among 60-cm tall herbs in a rocky spring trickle, still others perched close to water farther along same stream. Note that there is an undescribed species similar to Reventazon and in the same habitat but with more extensive green markings. In that species, base of antehumeral stripe is a clearly defined dash in both sexes, males have a black-tipped abdomen, and females have a dull yellow wash on the side of S9 and ovipositor as long as cerci. **Range:** Southern Caribbean slope, 500–900 m. **Costa Rican endemic.**

male

female

genus *Palaemnema* (p. 37)
male appendages in lateral view; left cercus in dorsal view

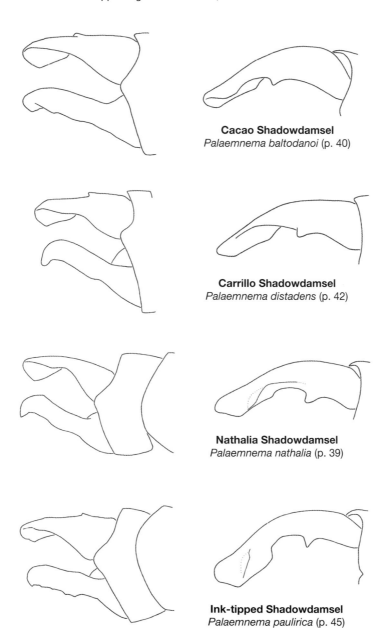

Cacao Shadowdamsel
Palaemnema baltodanoi (p. 40)

Carrillo Shadowdamsel
Palaemnema distadens (p. 42)

Nathalia Shadowdamsel
Palaemnema nathalia (p. 39)

Ink-tipped Shadowdamsel
Palaemnema paulirica (p. 45)

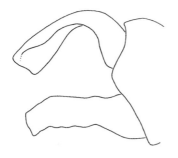

Arch-tipped Shadowdamsel
Palaemnema collaris (p. 38)

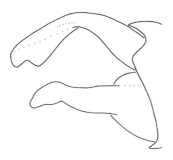
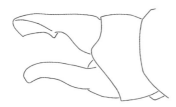

Chiriquita Shadowdamsel
Palaemnema chiriquita (p. 37)

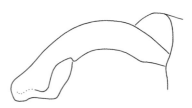

Dentate Shadowdamsel
Palaemnema dentata (p. 41)

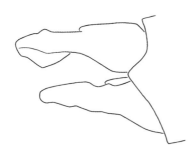

Janet's Shadowdamsel
Palaemnema joanetta (p. 44)

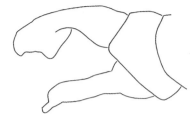
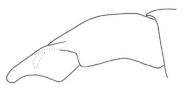

Elongate Shadowdamsel
Palaemnema gigantula (p. 43)

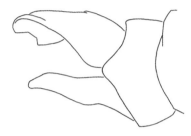
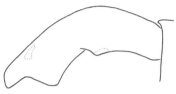

Reventazon Shadowdamsel
Palaemnema reventazoni (p. 46)

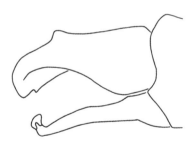

Black-backed Shadowdamsel
Palaemnema melanota (p. 39)

Broad-winged Damsels (family Calopterygidae)

Nine species. Members of this family occur throughout much of the world, although the species are especially diverse in the Asian and American tropics. They are among the largest and most beautiful of damselflies, characterized by broad, heavily veined wings that lack basal petioles. *Hetaerina* is the only genus in Central America. Most species live along forest streams, and both sexes are commonly found there; even immatures may stay near water. Males rapidly circle one another horizontally in flights of attrition, until the loser flies away. Females often descend underwater to oviposit. Males of all species are characterized by at least some red pigment on the bases of the wings; females have at most a red wash. Both sexes of River, Highland, and Dot-winged lack stigmas, as do some Smoky. Males of Bronze, Purplish, and Red-striped have very short, not visible, paraprocts. Males of River, Bronze, Purplish, Racket-tipped, and Red-striped have a red spot at the hindwing tips; Bronze and Racket-tipped have a red spot at the forewing tips as well. If spots are present in the others, they are smoky brown (tinged with reddish reflectance in some Highland). Female Dot-winged, Red-striped, and Forest have metallic red on the thorax; the other species have metallic green. Rubyspot species also differ in colors on the labrum and postclypeus, so close views of the face are helpful.

Highland Rubyspot   *Hetaerina cruentata*

TL 46 mm; HW 27 mm. **Identification:** Stigmas lacking. Mature males have green-striped thorax, brown wing tips (rarely with red tint), and paraprocts that are ¾ length of cerci. Bronze males are the only others in the group with metallic green on the thorax, but they are larger, with no apparent paraprocts. Females distinguished by combination of green-striped thorax, prominent green triangle on side between bases of forewing and hindwing, green or blue face with pale labrum, and lack of stigmas. Other species lacking stigmas (Dot-winged, River, and Smoky) occur primarily in lowland habitats, not at mid-elevation streams that Highland frequents. **Habitat and Behavior:** Small and medium-sized streams in open areas; also wide streams and rivers in forest with plenty of sunlight. Occasionally at ponds. Tolerates poor water quality more than other rubyspots. **Range:** Pacific slope, 400–1600 m; Caribbean slope, 500–2300 m. Highlands of northern Mexico to Suriname.

male

female

TL 50 mm; HW 30 mm. **Identification:** In mature males, note presence of stigmas, bright red coloration on all wing tips, thorax marked with green stripes, and rudimentary paraprocts. On the three other species with bright red wing tips (Purplish, Red-striped, River), red appears only on hindwings. Racket-tipped has smaller, red-brown spots on all wing tips; also note long paraprocts. Males of these four species are distinguished from male Bronze by dark metallic thorax with narrow pale stripes (on male Bronze, brownish thorax shows green stripes). Females, with green-striped thorax, green postclypeus, and dark, pale-sided labrum, most closely resemble female Racket-tipped (p. 55). Very difficult to distinguish the two in the field, but they rarely occur together, largely replacing one another altitudinally (look for males to make the identification). **Habitat and Behavior:** Small clear streams in forest. **Range:** Pacific slope, 300–1330 m; Caribbean slope, 600–1200 m. Highlands of central Mexico to Ecuador and Venezuela.

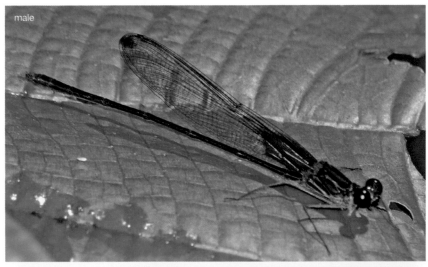

male

female

51

TL 58 mm; HW 35 mm. **Identification:** This montane species is the largest of CR rubyspots. Males have blackish (red glint in sunlight) thorax with pale stripes, red spot on hindwing tips, and rudimentary paraprocts, giving superficial impression of male Bronze (p. 51), but male Purplish is larger and lacks red spot on forewings. Also, male Purplish are the only rubyspots with abdomen metallic violet (when flying in sunlight, base of wings also show metallic violet). Highland Rubyspots (p. 50), which occur at similar elevations, are duller, with brown to reddish brown spots on wing tips and obvious paraprocts. Female Purplish are similar to some other species in having stigmas, green face, and dark, metallic green thorax with stripes, but note larger size and entirely dark labrum in Purplish. Pale markings on thorax are less extensive than in Bronze. Females also have notable red shading at wing base. **Habitat and Behavior:** Small to medium clear streams in forest. **Range:** Both slopes, 1100–2200 m. **Costa Rican endemic.**

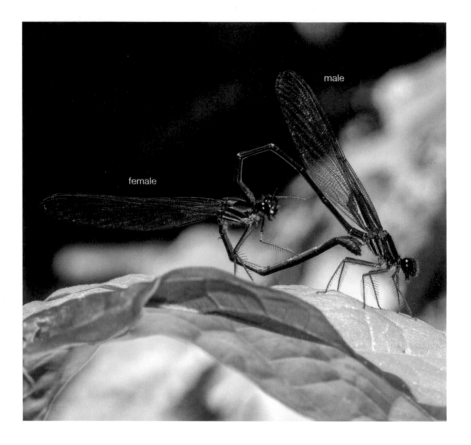

TL 44 mm; HW 25 mm. **Identification:** Males distinguished by combination of red spots at hindwing tips, metallic red thorax with cream stripes, red patch at hindwing base that extends into a stripe, presence of stigmas, and rudimentary paraprocts. Note that immature males lack the red wing tip spots. The only other species with red thorax and red spots on wing tips is River Rubyspot (p. 54), but River has prominent paraprocts and lacks stigmas. Female Red-striped can be identified by red- or black-striped thorax and red face; no other species has this combination. **Habitat and Behavior:** Small streams in forest; also found in streams passing through open areas with good shrub cover. **Range:** Caribbean slope, to 600 m in the Tilarán Mountain Range. Guatemala to Ecuador and Venezuela.

male

female

TL 48 mm; HW 28 mm. **Identification:** Males identified by combination of prominent red spot on hindwing tips, absence of stigmas, and paraprocts that are half length of cerci. Note that immature males sometimes appear at water, but they lack the red spots of mature males. Females characterized by combination of metallic green thoracic stripes and red to green face. Only other red-faced females, Red-striped (p. 53), have red- or black-striped thorax. Female Racket-tipped and Smoky (p. 58) also have green face and green thoracic stripes; note that Smoky, which shows regional variation, has dark brown wings where it occurs with River in Caribbean lowlands (Caño Negro region). Female Bronze and Highland also have green thoracic stripes, but neither overlaps with River. **Habitat and Behavior:** Clear streams and rivers in forest; also in ditches with rapid current. Along with Smoky, more likely to be on larger rivers than other CR rubyspots. **Range:** Pacific slope and inland to Caño Negro, to 200 m. Nicaragua to Peru and French Guiana.

male

female

TL 45 mm; HW 25 mm. **Identification:** Males distinguished by metallic red thorax, red-brown spots on all wing tips, presence of stigmas, and very long paraprocts with racket-like tips. Dot-winged (p. 56) shares red thorax and has brown wing tips but lacks stigmas, has much less red in wings, and has shorter paraprocts. With green-striped thorax, green or black face, and stigmas, females might be mistaken for Smoky, with which they often occur. See that species (p. 58) for differences. **Habitat and Behavior:** Small streams with clear water. **Range:** Pacific slope, to 600 m; Caribbean slope, 300–600 m. Northern Mexico to Peru and Venezuela.

male

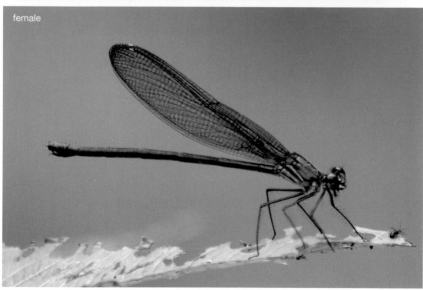

female

TL 50 mm; HW 28 mm. **Identification:** Only rubyspot with limited red in wings and no stigmas. Mature males distinguished by prominent brown spots on all wing tips and red, purple, or black thorax with reddish stripes. Racket-tipped (p. 55), which has a red thorax and brown to reddish spots on wing tips, has stigmas and very long, racket-shaped paraprocts. Female Dot-winged have metallic red thorax and black anteclypeus. Red-striped (p. 53), the only other females with red thorax, have stigmas and red face. **Habitat and Behavior:** Small forest streams, where it occurs with both Forest and Racket-tipped Rubyspots. **Range:** Southern Pacific lowlands north to Monteverde and southern Caribbean slope, to 400 m. Costa Rica to Ecuador.

TL 48 mm; HW 29 mm. **Identification:** Males have combination of metallic green labrum and metallic red thorax. This is also the only species in which mature males have a clear forewing tip and a hindwing tip with a brown spot. Hindwing spot larger and more diffuse than spots that occur on both wing tips in Racket-tipped and Dot-winged. Females, with red- or black-striped thorax, most similar to Dot-winged and Red-striped, but they lack stigmas and have green postclypeus and labrum (female Dot-winged are mostly black; Red-striped have red postclypeus). **Habitat and Behavior:** Small streams in forest. Common at tiny forest streams, especially where partly over-grown with shrubs and large herbs. Relatively poorly known in CR compared with other rubyspots. **Range:** Southern Pacific slope, to 500 m; Caribbean slope, to 1100 m. Belize to Colombia.

male

female

57

TL 47 mm; HW 28 mm. **Identification:** Individuals and populations vary. Only rubyspot males with significant amount of brown in wings; all with smoky wing tips. Also only ones with stigmas varying from well-developed to absent. Strongly upcurved paraprocts half length of cerci are also characteristic. Males of Caribbean populations have a black body, some with pale stripes low on sides of thorax, and show much black in wings. They vary substantially and apparently continuously from entirely dark wings to those with *less* color than is typical of Pacific populations. Brown on wing base of both wings typically reaches nodus or extends beyond on front edge. Even on darkest wings, some red veins are visible near wing bases. Pacific males usually have red thorax, with pale stripes always present. They have mostly red wing bases, with colored area of hindwing larger than in any other Costa Rican species, usually reaching to nodus or a little beyond. On females, thoracic markings metallic green to very dark. Females of Caribbean population often have entirely dark brown wings and are unmistakable, while wings of Pacific females not significantly darker than most other rubyspots. Pacific females very similar to female Racket-tipped (p. 55) but have paler face, with labrum and usually postclypeus pale. Female Racket-tipped have postclypeus always black and labrum ivory-white with black median stripe. **Habitat and Behavior:** Medium-sized streams to large rivers, usually in forests, but some occur in more open areas. **Range:** Lowlands of both slopes, to 300 m. Southern US to Panama.

Pacific

Caribbean

genus *Hetaerina* (p. 50)
male appendages in lateral and dorsal views

River Rubyspot
Hetaerina caja (p. 54)

Bronze Rubyspot
Hetaerina capitalis (p. 51)

Highland Rubyspot
Hetaerina cruentata (p. 50)

Dot-winged Rubyspot
Hetaerina fuscoguttata (p. 56)

Purplish Rubyspot
Hetaerina majuscula (p. 52)

Red-striped Rubyspot
Hetaerina miniata (p. 53)

Racket-tipped Rubyspot
Hetaerina occisa (p. 55)

Forest Rubyspot
Hetaerina sempronia (p. 57)

Smoky Rubyspot
Hetaerina titia (p. 58)

Bannerwings (family Polythoridae)

Four species. These damselflies have fairly broad wings with short petioles; the wings are clear, washed with color, or show dark tips of varying extent. Also note longish stigmas and many antenodal crossveins. Males have forcipate superior appendages with a tooth at mid-length, quite similar in all species. The tip of the female abdomen bulges with a large ovipositor. In all species the eyes are brown in both sexes. Coloration is extremely variable: the thorax is mainly black with fine stripes, mainly blue, or mainly yellow. In Variable Cora, males occur in four morphs; in Chirripo Cora, note dramatic altitudinal and perhaps geographic variation. Given the variation found in these four species, it is possible that a fifth species lies hidden, but more research is needed to settle the matter. The wings of the two most common species—Variable and Chirripo—may flash iridescent blue in the sun. All are restricted to forest streams, often perching in the shade. The larvae of all four species have gills along the abdominal segments, which is unusual among odonates and unique in CR. The terminal caudal lamellae are heavily sclerotized, presumably not functioning in respiration.

Blue Cora *Cora marina*

TL 46 mm; HW 27 mm. **Identification:** On males, abdomen is mostly blue; on females, brown to golden. On both, also note large size, long stigmas, and dense venation. Other coras all have black abdomens with pale colors only at base. Central area of wings darkens with age in males, as does basal area in females, but tips remain clear. **Habitat and Behavior:** Breeds in medium-sized rocky streams bordered by forest. **Range:** Pacific slope, 400–600 m; Caribbean slope, 600–1200 m. Southern Mexico to Colombia.

male

female

TL 40 mm; HW 25 mm. **Identification:** Could be a challenge to distinguish this from other cora species, especially *obscura* form of Variable (which has a striped thorax), but clear forewing tip and black spot at hindwing tip of both sexes of Peralta are distinctive (though must be seen up close). **Habitat and Behavior:** Small streams and spring trickles in forest. Often perches on dead twigs over water, even in shade. **Range:** Caribbean slope, 200–600 m. Costa Rica to Ecuador.

TL 47 mm; HW 32 mm. **Identification:** Individuals vary greatly, even within populations; some with clear wings, others with black wing tips. Thorax pattern also varies, from unstriped to heavily striped. On thorax, amount and hue of pale markings varies with age, usually changing from cream-yellow to sky-blue in mature males. Note that individuals from higher elevations are substantially larger than those from lower elevations. Males occur in at least four color forms, assuming all are the same species; the three most common are described here. Form *chirripa*: Labrum black and ivory; thorax blue with black median area of varying size; wing tips clear. Form *donnellyi*: Similar to *chirripa* but black median area wider and often uneven, and wing tips black. Blue may be replaced by cream-yellow in northern Guanacaste. Form *skinneri*: Labrum all black; thorax black with three pale lateral stripes; wing tips show more black than does *chirripa*. All forms distinguished from Blue Cora (p. 61) by black abdomen, from Variable (p. 64) by blue-sided prothorax, and from Peralta by having both wings either clear or tipped with black, although black in some individuals quite limited (Peralta has just hindwings tipped black). Some individuals, especially of form *skinneri*, show iridescent blue flash in wings, like Variable. Females not distinguishable from one another in the field. **Habitat and Behavior:** Small streams in forest. The most abundant and widespread of the cora species found on the upper slopes of the Monteverde region. Males perch on leaves and twigs near small streams and spring trickles in forested sites. **Range:** Generally on both slopes, 500–1600 m, rarely down to 100 m. Form *chirripa* is the most widespread, occurring on both slopes, 700–1600 m; form *skinneri* occurs mainly at higher elevations; form *donnellyi* occurs primarily on Caribbean slope. **Costa Rican endemic.**

TL 43 mm; HW 26 mm. **Identification:** A confusing species to say the least. Males of three color morphs are identical in structure, habitat, and behavior; females appear to be of a single type. Form *semiopaca*: Has striped thorax and broad black wing tips; distinctly larger than other two. Form *obscura*: Has striped thorax but note entirely clear wings. Form *notoxantha*: Thorax lacks stripes; most individuals have yellow thorax and yellow face, some have blue thorax and blue face (Monteverde area). Both *semiopaca* and *obscura* morphs show iridescent blue on wings in sunlight, latter perhaps only in younger individuals. Females have orange to yellow-brown thorax with black stripes and mostly black abdomen; wing tips almost entirely clear; may not be distinguishable from female Chirripo (p. 63) in the field, but the two species rarely occur together. **Habitat and Behavior:** Based on current information, appears restricted to Pacific slope. Small streams and spring trickles in forest. Males perch on leaves and twigs at edge of small forest streams (0.5–1 m wide). Females oviposit in rotten wood at water surface, sometimes with a guarding male following within 20–40 cm. Interestingly, form *notoxantha* tends to be warier than other two. **Range:** Pacific slope, to 1200 m. Costa Rica and Panama.

male

female

male
obscura type

male
notoxantha
blue type

male
notoxantha
yellow type

female

65

Flamboyant Flatwings (family Heteragrionidae)

Five species. This and the following family (Philogeniidae) formerly belonged to the family Megapodagrionidae, which biologists have concluded contains groups of damselflies that are more distantly related than previously thought. As the name *flatwing* suggests, these damsel-flies perch with the wings opened (even farther than those of spreadwings). Male *Heteragrion* are colorful, often mostly red or with vivid black and yellow patterns. Females, considerably shorter than males, are camouflaged with shades of brown and difficult to distinguish from one another. In all species, the upper parts of the eyes are dark brown to blackish, the lower parts greenish to yellowish; note that mature spreadwings usually have blue eyes. Both sexes perch with long abdomens hanging down at an angle. Males guard pools in small streams and spring trickles while perched on dead twigs or leaves near the water's edge; in face-to-face challenges, they hover with abdomen down at a 45° angle and fly at each other. Females perch on dead twigs in small gaps in the understory, some distance from streams; after mating, a female ovi-posits in tandem with a male or, less often, solo, frequently in soft wet wood.

Pale-faced Flatwing *Heteragrion albifrons*

TL 45 mm; HW 23 mm. **Identification:** Shiny ivory-yellow face distinguishes males from the only other red flatwing, Red-and-black. If face is not visible, note narrow dark abdominal rings (absent in Red-and-black). Females virtually indistinguishable from female Red-and-black unless up close; the two can be separated only by color pattern on top of head (see Red-and-black). **Habitat and Behavior:** Small forest streams and spring trickles. Common on Caribbean slope, in Monteverde area; rarely occurs at higher elevations on Pacific slope. Also found in wet forests of the Pacific southwest. Male's face, almost glowing, is prominent during male interactions in the shady forest understory. A tandem pair oviposited in a floating leaf lodged against a rock at the surface of a stream, male supported at 45° angle with wings fanning. **Range:** Pacific slope, to 1200 m; Caribbean slope, to 900 m. Much more common and widespread on Caribbean slope. Eastern Mexico to Costa Rica.

male

female

TL 45 mm; HW 23 mm. **Identification:** Combination of dark thorax and brilliant red abdomen is distinctive. Males distinguished from Pale-faced Flatwing by dark face and absence of narrow dark rings on abdomen. Females indistinguishable in the field from female Pale-faced; but up close, pattern on top of head in mature females differs. In Red-and-black, markings on top of head usually include black area that surrounds middle ocellus; Pale-faced usually have black crossbar in front of ocellus. Red-and-black females often have black labrum; in Pale-faced females, labrum is usually pale, contrasting with black postclypeus just above it. **Habitat and Behavior:** Small streams and spring trickles in forest; occasionally at isolated pools and ponds. Males guard pools and quieter places in small streams from perches on dead twigs over edge of water; they have been spotted on such perches even during heavy rains. Females perch on twigs in forest, 5–20 m from water, often at edge of small light gaps. Pairs remain in tandem after mating; tandem pair oviposited in floating leaf caught at surface in swift part of stream. **Range:** Both slopes to 1100 m; more common on Pacific slope than on Caribbean slope. Honduras to Ecuador.

male

female

TL 44 mm; HW 22 mm. There is little information about this species. **Identification:** Orange head, bordered at rear by black, is distinctive in males, as is clearly defined black median line on thorax. Females much like females of Pale-faced (p. 66) and Red-and-black (p. 67), but head shows more vivid black and yellow bands. Black posterior lobe of prothorax distinctive in both sexes. **Habitat and Behavior:** Small forest streams. **Range:** Caribbean slope, to 400 m. Costa Rica to western Colombia.

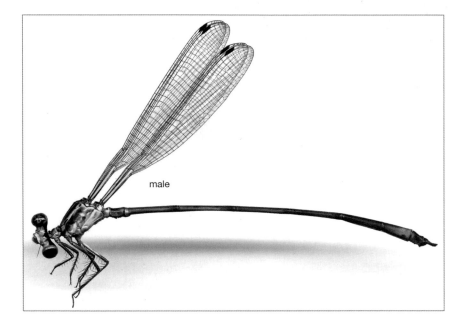
male

TL 43 mm; HW 22 mm. **Identification:** On both sexes, thorax shows vivid black and yellow stripes. Males have black head with labrum and postclypeus brilliant yellow, abdomen orange-brown banded with black. Females sometimes have yellow markings on top of mostly black head; abdomen darker than in other female flatwings, except for Mountain (p. 70), which is larger and has more yellow on abdomen tip; little overlap in elevation. **Habitat and Behavior:** Occurs along forest streams, especially intermittent or seasonal ones, and rarely around small ponds,. Males perch on twigs or rocks in sunny spots over running water. Both sexes perch on dead twigs 0.5–1.5 m above the ground, up to 10 m back away from water. Males in same spot over period of several days, presumably territorial. **Range:** Caribbean slope, 200–900 m. Costa Rica to Colombia and Venezuela.

male

female

TL 57 mm; HW 32 mm. **Identification:** Both sexes much larger than other flamboyant flatwings; in males, note striking yellow face, striped thorax and dark abdomen with bright yellow-orange tip. Females also distinguished by bright yellow tip on abdomen. This is the only flamboyant flatwing that is restricted to higher elevations. **Habitat and Behavior:** Small to medium streams in upland forest, also smaller side channels and quiet sections of larger rivers. Males guard quiet pools from perches above edge of pool and circle face-to-face with other males. Females oviposit in rotten wood at or just below water surface, usually in tandem with male. **Range:** Both slopes, 700–1600 m. Costa Rica and Panama.

male

female

Dusky Flatwings (family Philogeniidae)

Five species. Members of this family could be confused with spreadwings, as males perch with open wings and in most cases have a pruinose abdomen tip; they can be distinguished from spreadwings by their brown eyes and relatively shorter abdomen. Also note that they usually perch horizontally and thus could also be mistaken for a very slender clubtail (Gomphidae). The thorax is boldly patterned in younger individuals, showing brown on top with paired black median stripes and black and white stripes on the sides. With age, the top becomes darker and the sides obscured with pruinosity in both sexes. Pointed white lateral markings at the base of each abdominal segment also become obscured with age. Dusky flatwings often perch on tops of leaves. Most have dull coloration, though Blue Flatwing has a blue thorax and is colored much like Chirripo Cora. The other four species resemble each other, and males must be distinguished by their appendages. Females can be distinguished only by very slight differences in the prothorax. Most species are geographically separate, with only Costa Rican and Blue known to overlap, in Guanacaste NP and Monteverde.

Costa Rican Flatwing *Philogenia carrillica*

TL 55 mm; HW 36 mm. **Identification:** Males virtually identical to male Limon Flatwing (p. 73); even the appendages are similar, but cerci less acutely pointed in this species (see drawings for slight differences). Females difficult to distinguish even in hand. These two species overlap very little if at all; a few individuals from lower elevations in Braulio Carrillo NP are perhaps intermediate, but more specimens are needed to determine if the two species should be combined. **Habitat and Behavior:** Most common species on both slopes in northern CR. Small to medium streams in forests. Males at rocky streams up to 6 m wide, often at riffles. As well as leaves, they perch on upright dead twigs, partly sunken logs and rocks, or even on the ground. Immature males and females perch 5–50 cm high on dead twigs or sapling stems in forest understory, especially next to light spots or small gaps that receive morning sun. **Range:** Occurs in the lowlands of the northern Caribbean slope, from Puerto Viejo de Sarapiquí around to the Pacific slope of the Guanacaste and Tilarán mountain ranges, up to 1600 m. On Pacific slope, near Monteverde, occurs down to about 600 m. Nicaragua and Costa Rica.

male

71

female

TL 56 mm; HW 36 mm. **Identification:** Males differ from male Costa Rican Flatwing (p. 71) only in details of appendages; cerci more sharply pointed in lateral view. **Habitat and Behavior:** Small forest streams. Males guard sites along small streams and spring trickles, much like Costa Rican Flatwings. Females perch on dead twigs in forest understory distant from water and usually in shade. **Range:** Southern Caribbean slope north to Braulio Carrillo NP, 200–900 m. **Costa Rican endemic.**

male

female

Terraba Flatwing *Philogenia terraba*

TL 50 mm; HW 31 mm. **Identification:** Very similar to other *Philogenia* although slightly smaller. See illustrations of male appendages (p. 77). Perhaps no overlap with other dusky flatwings in CR, but closest in range to Golfo Dulce. **Habitat and Behavior:** Small streams in forest. Often perches at edge of forests and light gaps. **Range:** Central Pacific lowlands, from San Mateo south to San Isidro de El General, up to 900 m. **Costa Rican endemic**.

male

female

TL 55 mm; HW 32 mm. **Identification:** Similar to Terraba, the other Pacific-side species, but male cerci in lateral view broad at base and narrowing to pointed tip (in Terraba, male cerci somewhat arched in middle and not narrowing significantly to tip). In females, hind margin of pronotum rather squarish, not rounded as in Terraba. No overlap in range yet recorded. **Habitat and Behavior:** Territorial males perch on twigs or rocks over small spring outflows at the head of small forest streams. Other males and females perch close to the ground in forest understory well back from streams, often near small sunspots. Two males displayed to each other while perched side by side and facing in the same direction; one would leap 30 cm and then fall to its perch, and then the other would do the same. **Range:** Occurs on Pacific slope, in lowlands of Golfo Dulce and Osa Peninsula and up to San Vito, at 1200 m. Costa Rica and Panama.

male

female

TL 54 mm; HW 32 mm. **Identification:** The easiest *Philogenia* to identify. Males have broad black median stripe on blue thorax. In females, thorax is yellow-green in lower part, which is dark in other species. Superficially resembles Chirripo Cora (p. 63) but larger, perches with wings open, and never has black wing tips. **Habitat and Behavior:** Small to medium streams in forest. Locally common on streams and small rivers on northern Pacific and Caribbean slopes, from Monteverde to Guanacaste NP. Males perch on dead twigs and leaves along rocky forest streams. Females perch in understory 1–1.5 m above ground, 5–20 m back from water. **Range:** Mountain ranges in the northern region of the country, on both slopes, 900–1600 m. **Costa Rican endemic.**

male

female

genus *Philogenia* (p. 71)
male appendages in lateral and dorsal views

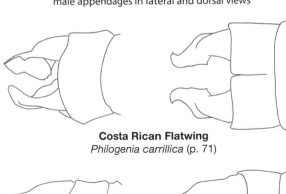

Costa Rican Flatwing
Philogenia carrillica (p. 71)

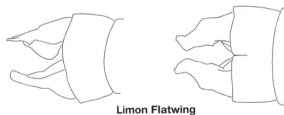

Golfo Dulce Flatwing
Philogenia championi (p. 75)

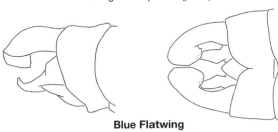

Limon Flatwing
Philogenia expansa (p. 73)

Blue Flatwing
Philogenia peacocki

Terraba Flatwing
Philogenia terraba (p. 74)

Cascade Damsels (family Thaumatoneuridae)

One species. Formerly included in Megapodagrionidae, this species is very large, with wings broadened in the middle; perched individuals hold their wings closed. The eyes are dark brown, unlike the blue eyes of large spreadwings, which might co-occur. The odd shape of the wing may be an adaptation to flying up and down the face of waterfalls. Species of the related genus *Paraphlebia* are smaller, have unmodified wings, and live farther north in Central America and Mexico. Male *Paraphlebia* that have black wing markings hold territories, while those with clear wings move up and down streams and attempt to mate with females in the territories of males with black wing markings. It is possible that the Great Cascade shows similar behavior.

Great Cascade Damsel   *Thaumatoneura inopinata*

TL 76 mm; HW 48 mm. **Identification:** Apart from the helicopters, this is the largest damselfly in CR, distinctly larger than next largest, the Cloudforest Spreadwing (p. 28), and distinguished from that species by thorax coloration—black with yellow lines—and wings held closed. Males either have black wing markings (wide central black band of variable extent) or clear wings. Females also polymorphic; in some, apical fourth of wing black, in others wings entirely clear. Even with two color morphs in each sex, not likely to be mistaken for anything else. **Habitat and Behavior:** Breeds at small to medium streams flowing over waterfalls in forested habitats. Males perch in sun, hanging down from branches and lianas, and fly back and forth along vertical banks with seepage. Both sexes also occur in forest light gaps well away from water sources, even in areas without waterfalls. **Range:** Caribbean slope, 700–1400 m; also on Pacific slope at Monteverde. Costa Rica and Panama.

male

male morph
with clear wings

female

79

Pond Damsels (family Coenagrionidae)

Sixty-seven species. Oddly, none of the members of this family has a common name that includes the word *pond*. This large family includes damselflies that will be familiar to naturalists from northern temperate zones. They are common in all aquatic habitats, from lowland marshes and streams to boggy montane ponds and seeps. A number of common US pond damsels occur all the way down to CR, including Dusky Dancer, Familiar Bluet (uplands), Citrine and Rambur's Forktails, and Desert Firetail. But the majority of species found in Costa Rica do not occur in the US.

The largest genus (*Argia*) comprises the dancers. These are almost all stream-dwellers, though a few species can be found at marshy seeps and ponds. Males of one group of species stand out with red eyes and a coppery thorax, while other *Argia* are mostly bright blue, mostly black with blue markings, or almost entirely dark. Those with mostly blue coloration are usually found in relatively open areas. In flight they appear to bounce up and down rather than fly in a straight line, and they hold their wings above their abdomen when perched. Wedgetails (*Acanthagrion*) have a blue-striped thorax and blue-tipped abdomen; in males, the end of the abdomen is a bit elevated because of the vertical expansion of the appendages and abdominal segment 10. Pearlwings (*Anisagrion*) have short pale forewing stigmas and change color rather dramatically with age; they are common at mid-elevation wetlands. Swampdamsels (*Leptobasis*) are slender, colorful damselflies that also change color greatly with age; they occur in forested lowland swamps. Firetails (*Telebasis*) are mostly bright red-orange and occur in a wide variety of pond habitats. Spinynecks (*Metaleptobasis*) are long and slender damsels; locally common, they are found in shaded swampy wetlands of lowland forests. Sprites (*Nehalennia*) and some forktails (*Ischnura*) are tiny and often go unnoticed in the vegetation bordering ponds.

This family includes 5 species of helicopter damsels, formerly considered a family of their own, Pseudostigmatidae. Genetic studies, as well as a few genera that bridge the gap between the two families, have shown the group to be members of Coenagrionidae. Although these damselflies don't fly anything like actual helicopters, they certainly bring them to mind. With four wings beating independently, the wing tips seem to whirl around these large species with their long abdomens. The smallest species is still longer than all other damsels in CR. Blue-winged is the most spectacular, with its blue-black-banded wings, and is iconic of the tropical rainforest.

The 7 species of threadtails (in three genera) were also considered a separate family, Protoneuridae, but genetic studies have shown that they also belong to Coenagrionidae. Most threadtails are small, very slender damselflies, although *Neoneura* are thicker-bodied, more like other pond damsels. The two Costa Rican *Neoneura* show a lot of red coloration. *Protoneura* are also colorful, with brilliant red, orange, or yellow markings. Females are duller, with the posterior half of the abdomen conspicuously enlarged. *Psaironeura* are the most inconspicuous, with dull orange on the thorax and a tiny area of pruinosity at the tip of the male abdomen. All hover or perch over slow-moving lowland streams, usually forested, though they seek out sunny patches when foraging.

DANCERS   genus *Argia*

Twenty-eight species. *Argia* is the largest genus of odonates in Costa Rica; it is more than twice the size of the next largest genera *Epigomphus* (knobtails) and *Micrathyria* (tropical dashers), with 13 species each. This very diverse New World genus is known to contain 139 species, and ongoing research indicates that many more are likely to be described. Most species breed in flowing water, from tiny trickles to wide rivers, though a few can be found at still waters. Dancers are rarely seen in copula but commonly in tandem, as pairs remain together for long periods of oviposition. This prolonged association facilitates the identification of females. Good oviposition sites may attract many pairs, even of more than one species.

Dancers are distinguished from most other pond damsels by relatively long tibial spurs (longer than the distance between them); wings that are held above the abdomen while perched; and a bouncing pattern in flight (thus the common name)—they seldom hover. Dancers often perch on rocks, logs, and bare ground, as well as on plant stems and leaves. Other similarly colored pond damsels (bluets, forktails) hold their wings alongside their abdomen, fly smoothly, often hovering, and almost always perch on vegetation. Tropical Sprites also have long tibial spurs and hold their wings above their abdomen, but they are smaller and slenderer than any dancer, with the sides of the thorax unstriped. Wedgetails hold their wings above their abdomen but are slenderer and have short tibial spurs.

Because they are so diverse, dancers present a real identification challenge, and some species can only be distinguished in the hand or with the aid of photos that show a close-up view so the appendages of a male and mesostigmal laminae of a female can be seen clearly. Upland species of open streams have a mostly blue abdomen; lowland species of forest streams usually an abdomen that is mostly black, with a blue tip. In four species, mature males have red eyes and a metallic thorax, and two species are mostly dark, with little or no blue. There is also great variation in size; the largest species are twice the bulk of the smallest.

To facilitate comparisons, the species are arranged in seven groups based on male similarities. The color pattern of the abdomen tip is important in distinguishing females; it can be entirely dark, dark with pale transversely oriented bands or spots ("banded"), with dark and light stripes ("striped"), or entirely pale. Note that females of at least some species are polymorphic, with pale markings either brown or blue, and a few combine those colors. Further complicating the situation is the pronounced geographic variation within Costa Rica of Oculate, Varied, Popoluca, and Black-fronted Dancers. Bear in mind that up to eight species can be found on a stream that provides the varied microhabitats that suit them.

Group 1: Males with red eyes and coppery thorax. Four species. Males in this group are distinguished by red eyes and a coppery thorax. Females have dull reddish eyes, a faint metallic red reflectance on the front of the thorax, and black stripes on the tip of the abdomen.

Ruby Dancer   *Argia cupraurea*

TL 37 mm; HW 22 mm. **Identification:** In Caribbean slope males, S7 is mostly blue; in Pacific males, S7 is mostly black (but occasionally blue). Overlaps with Fiery-eyed (p. 83) in Pacific lowlands, from Monteverde to Golfo Dulce; male Fiery-eyed has pale orange labrum (dark red metallic in Ruby). Male Garnet (p. 85) distinguished by mostly black abdomen. Male Calvert's (p. 84) very similar but a bit larger and more common in uplands. Females of all four species very similar, but on female Ruby note very short fork in humeral stripe (much longer in female Fiery-eyed). Ruby females all have brown thorax and blue abdomen tip; Fiery-eyed females are polymorphic, and some have blue thorax. Female Ruby also very similar to female Tezpi and Dusky, but broad humeral stripe shows coppery tinge rather than plain black. **Habitat and Behavior:** Found in both forested and open areas, at small to large rivers with shallow gravelly riffles. The common red-eyed dancer of wet lowlands on both Caribbean and southern Pacific slopes. More common on Pacific slope than Caribbean; can be abundant at some sites. **Range:** Pacific slope, from Monteverde south to Golfo Dulce, to 1200 m; Caribbean slope, to 700 m. Honduras to Venezuela.

male

female

TL 36 mm; HW 22 mm. **Identification:** Fiery-eyed overlaps with Ruby (p. 81) from Monteverde south, where males distinguished by pale orange labrum (dark red metallic in Ruby). Male Fiery-eyed has black occupying about ¼ to ⅓ length of middle abdominal segments (S3-6); on male Ruby and Calvert's (p. 84), black occupies no more than ⅕ length. Calvert's also slightly larger and occurs at higher elevations. Male Garnet (p. 85) has mostly black abdomen and may not overlap with Fiery-eyed. Female Fiery-eyed distinguished from female Ruby by more extensive pale coloration on face; note also more deeply divided humeral stripe. **Habitat and Behavior:** Generally prefers drier regions of the country. Common at riffles on clear, usually rocky, rivers; males perch mostly on rocks. More abundant on Pacific side and more common in lowlands than at higher elevations. Oviposition on rootlets and on leaves and plant debris caught among rocks in riffles. More tolerant of polluted water than other dancers. **Range:** Lowlands of Pacific slope and local at higher elevations, to 1200 m. Uncommon on northern Caribbean slope, inland to Fortuna. Southwestern US to Panama.

male

female

TL 44 mm; HW 26 mm. **Identification:** Largest of the red-eyed species. Greatly reduced black on brilliant blue abdomen distinguishes males from male Fiery-eyed (p. 83) and Garnet but not equally blue Caribbean-side Ruby. In lateral view, black stripe low on side of thorax better defined than in male Ruby (p. 81). On females in hand, note that mesostigmal laminae do not extend up into prominent lobes as they do in females of the other three red-eyes. The most common red-eyed species at middle elevations but not found in numbers comparable to Fiery-eyed and Ruby where they are common in the lowlands. **Habitat and Behavior:** Rocky, moderately swift forest streams. **Range:** Both slopes, 600–1300 m. Costa Rica and Panama.

male

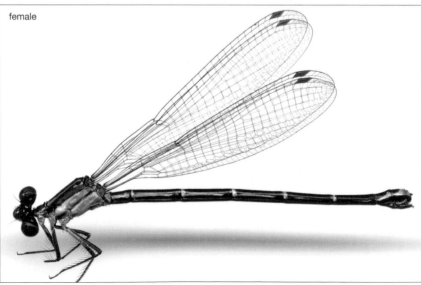
female

TL 39; HW 22 mm. **Identification:** Males the only red-eyed species in which abdomen is black through S7; S8 is blue for only half its length. Females indistinguishable in the field from female Ruby (p. 81). **Habitat and Behavior:** Much less common than other lowland red-eyed species in CR but occurs on forest streams like others, not on larger rivers. **Range:** Southern Pacific and Caribbean slopes to 700 m. Costa Rica to Ecuador.

Group 2: Males in which thorax and abdominal segments are mostly blue. Five species. Members of this group should be compared with the Familiar Bluet (p. 115) of highland lakes and ponds.

Cerulean Dancer *Argia anceps*

TL 39 mm; HW 23 mm. **Identification:** Males slightly paler blue than Thorn-tipped (p. 88), with which they often occur; no hint of purplish as in that species. Also note slightly amber-tinted wings (a good mark if seen in adequate light). Relatively wide median thoracic stripe distinguishes males from male Sky-blue (p. 90). Females share very narrow median stripe with female Sky-blue; female Thorn-tipped has wider median stripe. Both sexes essentially identical to Azure; see that species. **Habitat and Behavior:** In open areas, at a variety of types of running water, from spring trickles to small rivers; also at artificial ponds with minimal flow. Both sexes often perch on rocks or bare soil in sunny spots near water. A very common species. **Range:** Both slopes south to Central Valley, perhaps beyond, 500–1500 m. Southwestern US to Costa Rica.

TL 38 mm; HW 23 mm. **Identification:** Exactly the same as Cerulean Dancer in color pattern, both sexes of Azure differ in structural features. On male Azure, tori closer together (almost touching) than on male Cerulean (tori separated by almost their own width). On female Azure, lobes of mesostigmal plate much closer together than on female Cerulean (separated by nearly width of plate). Note the correlation between the parts of the two sexes that are engaged during tandem. The two species are at present not known to overlap in range, but note that few individuals have been collected in the area that separates their known ranges. **Habitat and Behavior:** Prefers small to medium-sized streams; less likely than Cerulean to be associated with ponds. **Range:** Pacific slope from San Isidro del General region south, 500–1200 m. Costa Rica to Colombia.

male

female

TL 36 mm; HW 22 mm. **Identification:** Of the dancers in this group, this species, Cerulean (p. 86), and Azure (p. 87) are all common at mid-elevation streams. Male Thorn-tipped differs from those two species by being slightly more purplish, having more black markings on sides of middle segments, and lacking even a hint of amber in wings. Also, lower end of humeral stripe has a black square larger than similar markings on Cerulean and Azure. Finally, long, pointed paraprocts assure identification of Thorn-tipped if seen up close (all other mostly blue species have short paraprocts). Also compare with Big Blue Dancer, larger and with forked humeral stripe. Abdomen of female Thorn-tipped pale—as in female Cerulean and Azure—but note pair of prominent black basal spots on S9. Female Cerulean rarely has similar basal spots and, if present, they are usually diffuse. Also, in female Cerulean the median thoracic stripe is almost nonexistent. **Habitat and Behavior:** One of most common and widespread dancers in CR. Lives in a variety of habitats, including streams and small rivers, drainage ditches, marshy spring seeps, and even ponds. Typically in open rather than forest. Like Cerulean, often seen perched on open ground. Oviposition usually in water with little or no current. **Range:** Both slopes, 500–1800 m. Eastern Mexico to Panama.

male

female

TL 45 mm; HW 28 mm. **Identification:** Males similar to other males in this group, but distinctly larger; of all the dancers in CR, this is the heaviest. Well-developed humeral stripe conspicuously forked; unforked in Thorn-tipped and Sky-blue (p. 90), forked in some Cerulean (p. 86) and Azure (p. 87). Abdomen has black markings similar to those of Thorn-tipped, with more black than Cerulean and Azure and considerably more than Sky-blue. Female has same forked humeral stripe and dark abdomen with banded tip (not pale as in Cerulean, Azure, and Thorn-tipped). Big Blue occurs with all other blue species at upper limits of their elevation range but also occurs well above that. **Habitat and Behavior:** Forested and open streams and small rivers, mostly above 1400 m. Oviposition on barely submerged rotten wood or well above water line on water-soaked logs; sometimes dozens of pairs at a favored spot. **Range:** Both slopes, 1100–2700 m. Highlands of central Mexico to Panama.

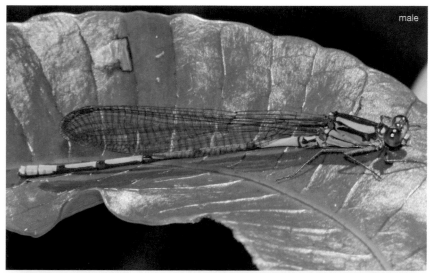

male

female

89

TL 37 mm; HW 23 mm. **Identification:** Males are bluest dancers in CR, with least amount of black on thorax and abdomen. Clearly smaller than Big Blue Dancer but similar in size to Cerulean, Azure, and Thorn-tipped, and it may occur with all of them. In addition to having black markings that are even more reduced, it is unique in that S10 is mostly black (blue in all others). Among this group of mostly blue species, female Sky-blue only one in which all pale markings are blue except for antehumeral stripes, which are brown; this contrast between brown top and blue sides of thorax also occurs in female Talamanca, Terira, and Waterfall Dancers (pp. 99–101), which also have black abdomen tip with S8 blue above. Of these, only Terira likely on same stream as Sky-blue. All blue S8 with half black S9 further distinguishes female Sky-blue from co-occurring mostly blue species. **Habitat and Behavior:** Favors clean springs, rocky streams, and rivers at mid-elevations, usually in more forested sites than Cerulean, Azure, or Thorn-tipped, though they often occur together. **Range:** Both slopes, 600–1700 m. Honduras to Ecuador and Venezuela.

male

female

Group 3: Males with thorax mostly black above; abdomen black, with blue rings or pointed dorsal markings; abdomen tip (S8-10 or S8-9) extensively blue. Nine species. This group should be compared with Neotropical Bluet (p. 116), of lowland streams, usually violet but varies to blue. Male wedgetails, which can also be on streams, are slenderer and have S10 black and somewhat elevated. Bluets and wedgetails also differ from dancers in flight style.

Oculate Dancer   *Argia oculata*

TL 35 mm; HW 21 mm. **Identification:** Shows great variation in CR. On Caribbean slope, males sky-blue with extended blue markings on S3-6. On Pacific slope, in Central Valley, and, uncommonly, on Caribbean slope (at La Selva, for example), males purple and black with only narrow rings of purple on S3-6. Geographic distribution of color forms is even more complex than this suggests—a blue population occurs on Nicoya Peninsula, for example. Females of both slopes have a banded abdomen tip. Males with blue markings are discussed in this group, but the species could also be included in Group 5. Males with blue markings are most easily mistaken for Varied (p. 92) or River (p. 95) Dancers. Male Varied Dancer, of similar size, is distinguished by black markings on S8; female Varied has distinctly longer ovipositor, extending prominently beyond abdomen tip and entirely black (tip pale in Oculate). River Dancer is a bit larger than Oculate. In male River, blue is a bit more extensive than in male Oculate; female River has a striped abdomen tip (banded in female Oculate). See also Olmec Dancer (p. 109) and Green-eyed Dancer (p. 106). **Habitat and Behavior:** Sunny areas along riffles in streams and rivers. One of the most abundant dancers in the lowlands of the Pacific slope but uncommon and local on Caribbean slope. Occurs on a wide variety of stream types, from slow and muddy to fast and rocky, sometimes at isolated rainpools. Blue color form may be found on smaller, more shaded streams than purple form. **Range:** Both slopes, to 1200 m. Northern Mexico to Bolivia and Brazil. (Further research may show that the two types of Oculate Dancer in Costa Rica are actually two different species, with a situation even more complex than described here.)

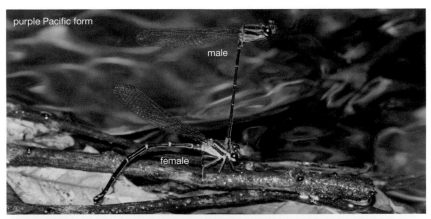

purple Pacific form

male

female

blue Caribbean form

female

male

TL 34 mm; HW 19 mm. **Identification:** Shows great variation. Superficially like Oculate Dancer (p. 91) but blue throughout its range; on Pacific slope, male Oculate is purple. In male Varied on Caribbean slope, middle abdominal segments have relatively small blue markings covering no more than half segment length; S8 has significant black coloration on sides that extends upward into the dorsal blue (sometimes entirely black). In some Caribbean-slope populations, antehumeral stripes relatively narrow and may be chartreuse. On Pacific slope, blue coloration on middle segments always more extensive than on Caribbean slope individuals, forming long triangles covering as much as half segment; S8 blue, with just a spot of black at rear. Wings usually amber-tinted on Caribbean slope but much less so on Pacific (good field mark when present). Females have thoracic striping much like Oculate and Olmec (p. 109), but ovipositor conspicuously larger, entirely black, and extending beyond tips of cerci. No other regional dancer has such a long ovipositor, the tips of which are almost always pale in other species. **Habitat and Behavior:** Favors small to medium forest streams with gentle current and either mud or rock bottom. **Range:** Both slopes, to 500 m. Honduras to Bolivia, Venezuela, and Brazil.

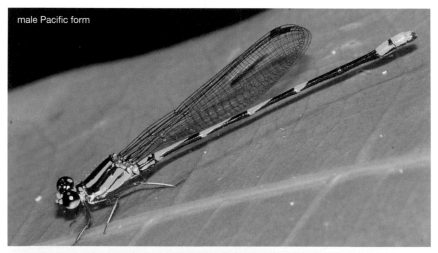

male Pacific form

female Pacific form

male
Caribbean
form

female
Caribbean
form

93

TL 38 mm; HW 23 mm. **Identification:** Males have blue-ringed abdomen and amber-tinted wings. Most similar to Varied (p. 92), which has black markings on S8. Several other species of similar size and markings have purple rather than blue on abdomen. Humeral stripe narrow and forked; other species with similar stripe have much more extensive blue on abdomen. Big Blue, Thorn-tipped, Sky-blue, and Waterfall Dancers occur with Mountain Spring; of them, only Waterfall (p. 100) with largely black abdomen. Female unknown. **Habitat and Behavior:** Only known individuals found at seeps and trickles in brushy pasture on forested mountain slope. **Range:** Known only from Pacific slope, Talamanca Mountain Range above San Isidro del General, 1700 m. **Costa Rican endemic**.

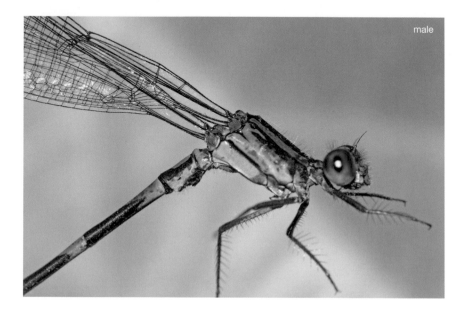

male

TL 41 mm; HW 21 mm. **Identification:** Males very similar to blue Caribbean-slope Oculate Dancer (p. 91), but larger and with relatively narrow humeral stripe notched on outside of upper end. Male Oculate generally has broader humeral stripe with notch within upper end of stripe. Males very distinct from purple Pacific-slope Oculate. Females have striped abdomen tip, unlike females of any species with similar males, and can be distinguished from other co-occurring females with striped abdomen tip (Fiery-eyed, Ruby, Tezpi, Dusky) by narrow, sometimes indistinct, humeral stripe. **Habitat and Behavior:** Rocky streams and rivers with moderate to fast currents in forested landscapes. Perches in sun on rocks, leaves, or dead twigs. Pairs oviposit in plants growing from rocks. **Range:** Both slopes, below 500 m. Known mostly on Pacific slope, from Guanacaste south to Cañas area. On Caribbean slope, single records from Alajuela, Heredia, and Limón (on Parismina River) provinces, so presumably widespread on Caribbean slope too. Nicaragua to Brazil but yet to be found in southern Costa Rica or Panama.

male

male

female

95

TL 32 mm; HW 19 mm. **Identification:** Two color forms. On Caribbean slope, most males have much blue on abdomen, including S8-10, like a miniature Oculate Dancer (p. 91) but with extensive blue on middle segments back through S6; in Oculate in the same region, the blue markings become smaller to the rear and by S6 occupy less than half segment length. On Osa Peninsula, males have less blue on abdomen, with middle segments almost entirely black, S8 with posterolateral black markings, and S10 in some all black. Black in Osa males also more metallic blue-green than in Caribbean males. In females of both populations, abdomen has black tip; on Caribbean slope (perhaps elsewhere), females have pair of pale posterior spots on S8 that fade with age. Note that both color forms occur in Braulio Carrillo NP and perhaps elsewhere. Could this be another case of two sibling species? **Habitat and Behavior:** Sunny areas in forest, at spring seeps and small streams. **Range:** Pacific slope, known only from Osa Peninsula; Caribbean slope, to 900 m. Eastern Mexico to Ecuador.

male Pacific form

male Caribbean form

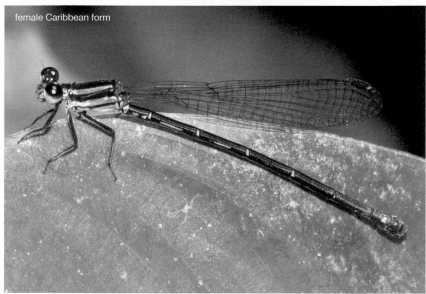

female Caribbean form

TL 34 mm; HW 19 mm. **Identification:** Males have narrow blue rings on abdomen and blue S8-10; very similar to male Popoluca Dancer (p. 96) but Popoluca on Pacific slope has black at tip of S8. Females of both species also similar; in female Golfo Dulce, note blue markings on S8 and S9, as well as smaller ovipositor (on Pacific slope, female Popoluca has abdomen tip entirely black). Male superficially similar to male Oculate Dancer (p. 91), but Oculate individuals in Golfo Dulce area are purple. **Habitat and Behavior:** Small streams and trickles within forest appear to be typical habitat, but poorly known. **Range:** Pacific slope, Uvita south to Golfo Dulce, to 300 m. **Costa Rican endemic**.

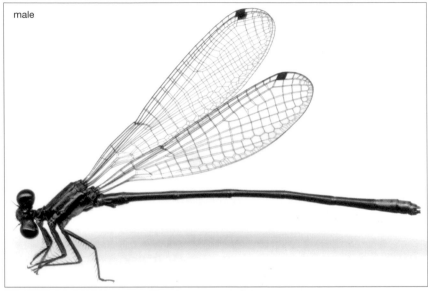

male

female

TL 34 mm; HW 21 mm. **Identification:** Males recognized by almost entirely blue thorax with narrow black median stripe (may also have very narrow black lateral stripes) and mostly black abdomen with only S1-3 and S8-9 blue above. Females have only S8 blue above. Female very similar to female Waterfall Dancer (p. 100), difficult to distinguish in the field. Small postocular spots of female Talamanca do not reach eye margins, while larger, oblong spots in Waterfall reach eye margins. Some females have blue longitudinal markings on top of S9, lacking in most Waterfall. **Habitat and Behavior:** Small streams in forest, especially where water flows down rock walls. **Range:** Southern Pacific and Caribbean slopes, to 900 m. Costa Rica to Ecuador and Venezuela.

male

female

TL 38 mm; HW 23 mm. **Identification:** A relatively large species. Males characterized by purple antehumeral stripe (but blue in some males) that contrasts with blue sides of thorax; abdomen black overall with most of S3 and all of S8-9 blue. Females also distinctive, with brown antehumeral stripe and blue sides of thorax (but some have brown sides); abdomen black with S8 blue and S9 black above. Probably not distinguishable from female Talamanca (p. 99) except up close; in female Waterfall, black humeral stripe full width at upper end, narrow at upper end in female Talamanca. **Habitat and Behavior:** Tiny spring trickles, rivulets, and waterfalls running down steep banks to a stream. Larvae crawl over mossy substrates on rocks and soil moistened by flowing water. **Range:** Both slopes, more common in cloud forest, 600–1700 m. **Costa Rican endemic.**

TL 43 mm; HW 26 mm. **Identification:** A large dancer of the highlands. Males distinguished by black abdomen, with half of S3 ringed with blue and S8-10 blue; also note amber-tinted wings. Another large montane species, Big Blue (p. 89), has humeral stripe forked but has a mostly blue abdomen. Female Terira have extensive blue surrounding abdomen base, a banded tip, and thorax that is brown and blue; also note blue lateral streaks on S3–6. Immatures of both sexes have distinctive conspicuous yellow stripe on side of thorax. Females similar to Pocomana (p. 110), but latter lacks black markings of Terira and has S3 with ¾ dorsal blue stripe and S9 all blue. Female Sky-blue (p. 90) smaller and lacks lateral abdominal streaks. **Habitat and Behavior:** Small rocky streams in cloud forest; also in semi-open areas at higher elevations. May be at same sites as Big Blue Dancer but tends to choose smaller and more heavily forested streams; sometimes at isolated pools. **Range:** Both slopes, 1000–2700 m. Costa Rica and western Panama.

Blue-gray Dancer *Argia carolus*

TL 35 mm; HW 22 mm. **Identification:** Males have pale coloration on head and thorax a dull blue-gray, sometimes with a greenish tinge, an unusual shade not seen in other dancers. Only S9-10 blue, but in some individuals there is a blue or purple horizontal line on either side of S8, a character seen in no other species. Underside of eyes also blue, contrasting with dull color of head and thorax. Females have similar coloration, abdomen tip banded with either blue or purple; could be confused with female Olmec (p. 109) or Oculate (p. 91) Dancers. **Habitat and Behavior:** Small forest streams, small flood channels and isolated flood pools along larger rocky rivers. **Range:** Pacific slope, from Monteverde south to Osa Peninsula, to 600 m. **Costa Rican endemic.**

male

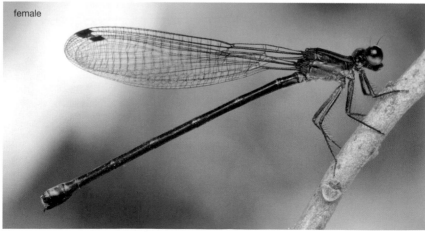

female

TL 29 mm; HW 17 mm. **Identification:** Smallest of CR dancers. Males show blue only on S9. Antehumeral stripes purplish or gray, contrasting with blue on sides of thorax. Basal abdominal markings also purple. Females brown or blue; andromorph has same contrast between antehumeral stripes and thorax as on male; also note abdomen with band across S8 and S9. Sometimes found with Popoluca (p. 96), which is also small, Bristle-tipped (p. 108), and Green-eyed (p. 140) Dancers. **Habitat and Behavior:** Found in forests and open areas, in swampy parts of small to medium streams, swampy spring seeps, and even extremely small trickles and isolated stagnant pools. **Range:** Both slopes, to 1300 m. Eastern Mexico to Bolivia and Brazil.

male

male

female

TL 40 mm; HW 23 mm. **Identification:** Males occur in two color forms: individuals on Caribbean slope are darker than those on Pacific, showing less blue; both lack postocular spots. Caribbean males almost entirely black above. Thorax has faint metallic tint like that of red-eyed species when seen in bright light. Small markings at abdomen base, tip of S8, and S9-10 blue. Locally common on Caribbean side but individuals have also turned up on northern Pacific slope near Monteverde where wet Caribbean climate spills over Continental Divide. Pacific males have some blue on face and pale blue on thorax, including antehumeral stripe and blue triangles at base of S3–6; S7 all black. Absence of postocular spots, as well as black anteclypeus and labrum, distinguishes them from other dancers such as Oculate (p. 91) with blue-and-black striped thorax and blue abdomen tip. Females also lack postocular spots but have narrow pale antehumeral stripe; also note pair of dorsolateral pale spots on posterior part of S8 and S9 that become obscured with age. Those female Dusky (p. 111) that lack stripes at abdomen tip look very similar but have small postocular spots; the two unlikely to overlap. **Habitat and Behavior:** Small forest streams, in areas of gentle flow and riffles. Males perch in sun on leaves or twigs near water's edge, up to 1.5 m above water. **Range:** Pacific slope, 400–1100 m; Caribbean slope, 500–1400 m. Costa Rica and Panama.

male Caribbean form

male Pacific form

female Pacific form

Green-eyed Dancer   *Argia frequentula*

TL 34 mm; HW 19 mm. **Identification:** Mature males distinguished from male Purple Dancer by eye color, black above and green below, and absence on abdomen of tan ventrolateral streaks. In immature males, pale markings brighter than on mature males, and lower thorax and abdomen tip often sky-blue; as a result, immatures could be confused with several species that are blue when mature. Both sexes can also be confused with Bristle-tipped (p. 108), which often co-occurs; males of that species have distinctive paraprocts that are long and pointed. In male Bristle-tipped, purple on top of thorax is usually replaced by blue on lower sides, producing two-toned effect; note, however, that some Green-eyed individuals have same coloration. Both sexes of Green-eyed have humeral stripe split at upper end, with lower branch shorter; in Bristle-tipped, both branches reach upper end, leaving pale triangle between them. Pale markings encroach from below on black middle abdominal segments of Bristle-tipped, rarely visible in Green-eyed. Black markings low on S8 and S9 of female Bristle-tipped curve upward to rear; the markings remain level in female Green-eyed. **Habitat and Behavior:** In both open and forested areas, found in small to medium streams. A common damselfly on Caribbean slope and in wet forests of southern Pacific slope but much more local in dry forest of northern Pacific. Rarely found at same sites as similar Purple Dancer. **Range:** Both slopes, to 700 m. Central Mexico to Panama.

male

female

TL 32 mm; HW 17 mm. **Identification:** Eyes of mature male Purple entirely dark blue to purple, while in very similar Green-eyed Dancer eyes are black above and green below. Both sexes usually can be distinguished from Green-eyed by tan ventrolateral streaks along abdomen (abdomen looks entirely black in mature Green-eyed). Female Purple also has broader humeral stripe; in female Green-eyed, stripe narrower and with shorter outer branch. Note that these two are rarely found together, as they prefer different habitats. On Pacific slope, larger male Oculate Dancers (p. 91) also purple, looking very much like Purple and Green-eyed Dancers. Eyes of male Oculate mostly black, with small amount of blue below, unlike Purple. Also, large postocular spots almost connected by pale occipital bar between them in Purple and Green-eyed, no such bar in Oculate. Female Purple and Green-eyed with pale abdomen tip, Oculate with banded abdomen tip. **Habitat and Behavior:** More in open areas than in forests. One of most common and widespread dancers in CR, living in great range of habitats including clear rocky streams and rivers, isolated stagnant ponds, flood pools, and polluted ditches. **Range:** Both slopes, to 1000 m. Northern Mexico to Peru and Brazil.

male

female

TL 31 mm; HW 18 mm. **Identification:** Males easily confused with male Green-eyed Dancers (p. 106), which also have eyes black above; male Purple Dancers (p. 107) have mostly blue to purple eyes. Male thorax usually paler blue than thorax of male Green-eyed or Purple; also note elongate, pointed paraprocts (easy to see with hand lens and even binoculars). Females andromorphic; they look all blue and can be distinguished by pale abdomen tip with black ventrolateral markings curving upward at rear of S8 and S9. **Habitat and Behavior:** Small streams and marshy seeps in open or forested areas, sometimes heavily vegetated ponds. Often occurs with Green-eyed and Swamp, less often with Purple; can be locally abundant. **Range:** Caribbean slope, to 1200 m, extending into Pacific drainage near northern border (Guanacaste NP). Honduras to Panama.

male

female

TL 37 mm; HW 22 mm. **Identification:** In males, purple ground coloration can cause confusion with purple form of male Oculate Dancer (p. 91), but note that Oculate is slightly smaller. Although both occur on Pacific and Caribbean slopes, Olmec is rare on Pacific, where Oculate is much more common. In male Olmec, pale antehumeral stripe narrows upward toward wing bases; in Oculate, the stripe narrows considerably less, the edges parallel in most individuals. In hand, look at male cerci, which are clearly forked in Olmec, with longer inner branch; in Oculate, inner branch forms a downward-pointing tooth. Also, male Olmec has four antenodal cells in hindwing, while Oculate usually has three. Females can be distinguished mostly by larger size, but also because ovipositor projects just beyond abdomen tip (just to it in female Oculate). **Habitat and Behavior:** Small to large streams in forest. **Range:** Both slopes, to 1200 m; most common from 500 to 900 m. Northern Mexico to Colombia.

male

female

Pocomana Dancer   *Argia pocomana*

TL 35 mm; HW 21 mm. **Identification:** Males only CR dancers with mostly purple abdomen, although coloration varies from dull purple-blue to purple-green. Females distinguished by mostly dark abdomen and S8-10 purplish above, black below. Females of three slightly smaller species (Purple, Green-eyed, and Bristle-tipped, pp. 106-108) have S8–10 blue above (this can appear purple under some light conditions); note that in immature Pocomana, S8–10 may also appear blue. Female Purple and Green-eyed have much less black below on S8–10 than does female Pocomana; female Bristle-tipped has black lower edge of segments curved upward at rear of each segment. In addition, two black markings at base of S9 in female Pocomana are lacking in other females. Finally, humeral stripes in female Pocomana average narrower than in other three. In females, S3-4 or 3–5 shows a long purple dorsal stripe, and S3–6 distinctive brown lateral streaks. **Habitat and Behavior:** Found at seeps and small trickles from springs, usually in sunny areas. Only locally common, especially where clean water runs down rock faces creating trickling waterfall effect and sometimes at micro-waterfall sites in larger streams. **Range:** Both slopes, to 1100 m. Guatemala to Peru.

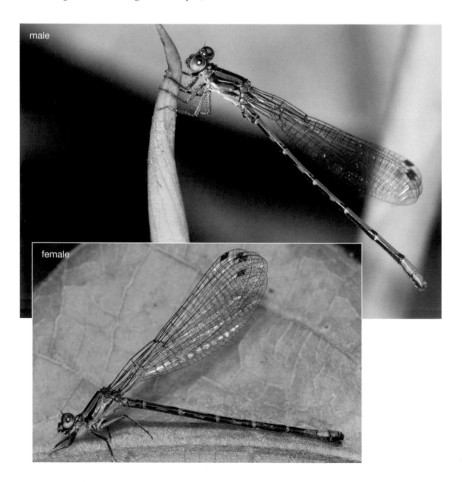

male

female

110

Dusky Dancer   *Argia translata*

TL 35 mm; HW 21 mm. **Identification:** Dusky and Tezpi Dancers are relatively long and slender, males with mostly black abdomen; wings almost clear in Dusky and amber-shaded in Tezpi (p. 113). When visible, best field mark is eye color: Dusky has blue to purple eyes and Tezpi has dark brown eyes. If visible, small postocular spots blue in Dusky and tan in Tezpi. Note that male Dusky, when mating, may undergo a dramatic transformation; the thorax sometimes becomes pale with dark stripes, as in female and immature male; male returns to normal coloration thereafter. Significance of this change unknown. Female Dusky very similar to female Tezpi, both with abdomen tips striped; may not be distinguishable in the field unless wing color is visible. Note, however, that in female Dusky stripes extend onto top of S10, which is black; S10 pale in female Tezpi. Of all species with stripe-tipped females, Dusky only one in which black on S9 covers entire top of segment, some females with abdomen tip entirely black. Finally, many female Tezpi have blue thorax; female Dusky rarely have blue thorax and perhaps never in CR. Upper part of humeral stripe in female Dusky with pale triangular streak, female Tezpi with small spot. **Habitat and Behavior:** Sunny areas along riffles in streams and rivers, rocky or not. Characteristic of larger rivers, where pairs oviposit on surface mats of leaves and plant debris. Both sexes perch higher in trees than do other dancers, often well away from water. **Range:** Both slopes, to 400 m. Extreme southeastern Canada to Argentina and French Guiana.

male

female

TL 38 mm; HW 24 mm. **Identification:** Slightly larger than Dusky (p. 111), both sexes with amber-tinted wings. Males of both have almost entirely black abdomen. Tip of S9 and S10 often with small blue markings (rarely entirely black); blue markings if present in Dusky would be on base of S9. Front of thorax of male Tezpi reflects metallic purple in bright light, while Dusky is matte-black. As in Dusky, males in tandem may show bright thoracic pattern, black above and whitish below, with pale, narrow antehumeral stripe. Females polymorphic, with brown or blue thorax (blue females quite rare in Dusky). Black markings on top of S9 extend to rear of segment in Dusky, not so in Tezpi. Female Fiery-eyed Dancer (p. 83) very similar, with striped abdomen tip, but wings not amber-shaded; the two occur together commonly in Guanacaste. In fact, females of all species of Group 1 (p. 81) have striped abdomen tips and look much like Tezpi. **Habitat and Behavior:** Sunny areas along rocky streams and rivers with abundant riffles, less often in slow muddy trickles. Oviposition in aquatic vegetation. Females often perch in vegetation at woodland edge. **Range:** Pacific slope to 700 m, not known from Golfo Dulce region. Southwestern US to Costa Rica.

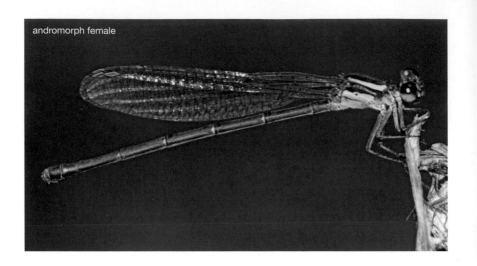
andromorph female

BLUETS   genus *Enallagma*

Two species. Although this is the largest genus of damselflies in temperate North America, only two very widespread species occur in Costa Rica. Bluets can be distinguished from dancers by their habit of holding their wings beside the abdomen (in dancers, above the abdomen) while perched and by their smooth flight style, often appearing to float just above the water surface (dancers fly erratically). See also wedgetails (p. 124) and pearlwings (p. 127).

Familiar Bluet   *Enallagma civile*

TL 33 mm; HW 19 mm. **Identification:** Bright blue males could be confused with male dancers that are mainly blue (Cerulean, Azure, Thorn-tipped, Sky-blue), but wing position when perched, smooth style of flight, and black S10 identify them as bluets (of the dancers, Sky-blue also has black S10). Females polymorphic; pale areas either tan to greenish (more common) or blue. Generally, females more closely resemble other female damselflies than males resemble other males, but smooth-edged thoracic stripes in bluets (usually a bit irregular in dancers) and behavior are good clues. Female forktails and firetails lack black humeral stripe. Female wedgetails have blue-tipped abdomen and are slenderer. Male pearlwings have more black on abdomen, while females have subdued thoracic striping. **Habitat and Behavior:** Open ponds, both natural and artificial, in uplands. Locally, may occur in great numbers, at higher density than other CR damselflies. Males patrol over open water, close to surface, while male dancers fly between perches. Tandem pairs seek out emergent and floating vegetation as oviposition sites; female may then submerge by herself to continue ovipositing. **Range:** Both slopes and Central Valley, 750–1900 meters. Southern Canada to Venezuela, West Indies.

male

female

115

TL 32 mm; HW 18 mm. **Identification:** Males are black and violet, with long, pointed cerci; they resemble some male dancers that are also violet, but they perch with wings beside the abdomen and fly smoothly. Also, in male dancers that are violet (Green-eyed, Swamp, Bristle-tipped, Oculate, Popoluca, Purple, and Olmec), S10 is always similar in coloration to S9, not black as in this species, and none has long, pointed cerci. Some male Neotropicals show blue low on thorax and at abdomen tip, but all males have pale color at abdomen tip interrupted by black ring at end of S8. Female polymorphic; thorax light brown to greenish or violet to blue. Female has distinctive pattern at abdomen tip, with S8 pale above on basal two-thirds or more and pair of pale spots on S9. **Habitat and Behavior:** Streams and small rivers, usually at pools rather than rapids. Males perch on stems with abdomen extending out noticeably (dancers almost never perch like that) or patrol back and forth close to water. Pairs oviposit in tandem on floating vegetation. **Range:** Both slopes, to 600 m; on Pacific slope, uncommonly to 1100 m. Southwestern US to Argentina and Brazil.

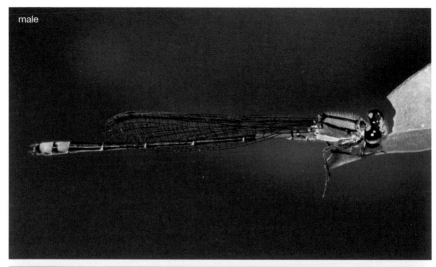
male

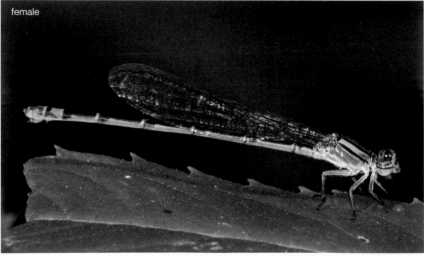
female

TL 30 mm; HW 15 mm. **Identification:** Males unmistakable, with yellow face, chartreuse or blue thoracic stripes, and blue-tipped abdomen. Long, black angular cerci add distinction to abdomen tip. Females polymorphic, with thorax either blue or greenish and somewhat similar to female bluets, but distinguished by abdomen tip (S8–10) irregularly marked blue and black (most individuals have more blue than female in photo). Female most resembles female Neotropical Bluet of stream habitats, but S8 mostly pale in Neotropical. **Habitat and Behavior:** Open ponds and large slow-flowing rivers; often flies over open water, perching on isolated emergent or floating plants. Females oviposit in tandem or alone, on floating vegetation. **Range:** Both slopes, to 300 m. Texas, Florida, and West Indies to Venezuela.

FORKTAILS genus *Ischnura*

Three species. Forktails typically occur in open marshes, where they are often found in dense grass and sedges. In all males, note a short, forked structure at the end of S10. Variation within a species is great in this genus, and coloration of immature females is dramatically different from that of mature females. Tiny and Citrine are among the smallest of Costa Rican damselflies (Tropical Sprite and Striped Firetail are also very small).

Rambur's Forktail *Ischnura ramburii*

TL 32 mm; HW 18 mm. **Identification:** Males unmistakable: green or blue (in immature) thorax shows black stripes, and black abdomen has blue tip (S8-9). Other damselflies with striped thorax and blue-tipped black abdomen are Tiny Forktail, which is much smaller; Caribbean Yellowface (p. 117), distinguished by its yellow face; wedgetails, which are more slender; and some dancers, all of which perch with wings above abdomen. Tropical Sprite (p. 148) is much smaller, with less obviously striped thorax and S8–10 all blue. Female Rambur's is polymorphic; andromorph female looks just like male but always with blue thorax. Heteromorph female distinctive, with either bright orange (immature) or greenish or brown (mature) thorax with black median stripe, and black abdomen. **Habitat and Behavior:** Breeds in open ponds with abundant emergent or bordering vegetation. Perches on emergent stems and lily pads. Tolerant of poor water quality, including brackish ponds. Both immature and mature females mate, then oviposit solo. Rarely seen away from water, unlike many other pond damsels. **Range:** Pacific slope, to 1400 m; Caribbean slope, to 800 m. Southern US to Ecuador and Brazil, also West Indies.

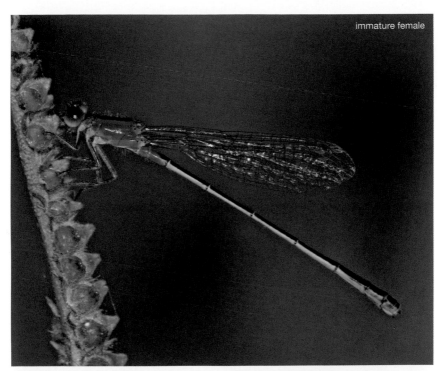

immature female

andromorph female

119

TL 22 mm; HW 11 mm. **Identification:** Very small size, green-and-black striped thorax, and blue-tipped abdomen are unique. Similarly marked male Rambur's Forktail (p. 118) is considerably larger, with S8-9 entirely blue (much black on S8 in Tiny). In andromorph females, thorax is blue-and-black striped. Immature heteromorph females are unmistakable, with entirely lemon or lime thorax. Mature heteromorph females, with dull greenish black-striped thorax and blue postocular spots (some have blue abdomen tip), look like small version of female wedgetail, bluet, or yellowface, and are best distinguished by size alone. **Habitat and Behavior:** Ponds and marshes with abundant emergent vegetation. Has been reported ovipositing in tandem, but that may not be normal mode. Usually not seen away from water, but clearly adept at dispersing, as they appear at newly formed wetlands. **Range:** Pacific slope and Central Valley, to 1400 m; Caribbean slope, to 900 m. Northern Mexico to Argentina, West Indies.

male

immature female

male

female

andromorph female

TL 24 mm; HW 12 mm. **Identification:** Very small. Males unmistakable, with black and green striped thorax and yellow abdomen; in immature female Tiny Forktail (p. 120), note yellow thorax and black abdomen. Commonly seen immature female Citrine is also distinctive; orange on most of body, with black mostly restricted to S7-8. Female Rambur's (p. 118) also has an orange thorax, but has abdomen black except at base and tip; also perceptibly larger. Red-orange firetails and orange immature pearlwings and swampdamsels all show different patterns. In mature female Citrine, note contrast between black median stripe and paler sides of thorax, whether brown, greenish, or pruinose, and highly pruinose gray to dull black abdomen. Mature female Citrine lose colored postocular spots with age, while larger heteromorph female Rambur's retain small blue spots. **Habitat and Behavior:** Marshy ponds, typically in dense grass and sedges. Females mate while in both immature and mature color stages. Ovipositing females especially difficult to find. **Range:** Pacific slope and Central Valley, to 1500 m; Caribbean slope, to 600 m. US to West Indies, Guianas, Colombia, and Galapagos.

male

female

immature female

123

WEDGETAILS genus *Acanthagrion*

Three species. Males of these slender damselflies perch on vegetation over water in grassy and sedgy areas. The end of S10 and the cerci are elevated to form a "wedge" at the end of the abdomen, varying in height among species. In males, eyes are black above and green or blue below; in females, eyes are brown above and tan or greenish below. With striped thorax and blue-tipped abdomen, males could be mistaken for bluets or forktails. Unlike those others, however, they hold their wings above their abdomen when perched, similar to dancers. But wedgetails are more slender, with S10 black instead of blue, and have larger postocular spots than those dancers with a mostly black abdomen. Females are duller but with patterns much like those of males. Oviposition is probably in tandem for all species, but mating behavior is poorly known in wedgetails.

Narrow-tipped Wedgetail *Acanthagrion inexpectum*

TL 32 mm; HW 17 mm. **Identification:** Males differ from other wedgetails by having extensive blue coloration on abdomen base as well as blue on tip of S7. Females distinguished from other slender species such as bluets by having blue only on abdomen tip; almost exactly like other wedgetails but have thoracic stripes that are less clearly defined. **Habitat and Behavior:** Marshy ponds. **Range:** Southern Pacific slope, to 1500 m. Eastern Mexico to Panama.

male

male

female

TL 32 mm; HW 17 mm. **Identification:** Males more likely to have green on thorax than male Pacific Wedgetail (p. 126), either entire pale area green or just antehumeral stripe. Up close, male S10 is noticeably lower than that of male Pacific. See Pacific for comparison of females. **Habitat and Behavior:** Usually in sun at ponds with abundant vegetation, both in open areas and in forest. Perches on lily pads and emergent grasses. Males "dance," facing each other 10–15 cm apart; as they rise slowly, they bounce up and down 1 or 2 centimeters, with the abdomen tip slightly bent upward in display. **Range:** Caribbean slope, to 1200 m. Nicaragua and Costa Rica.

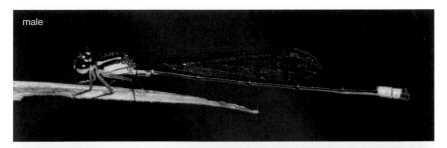

male

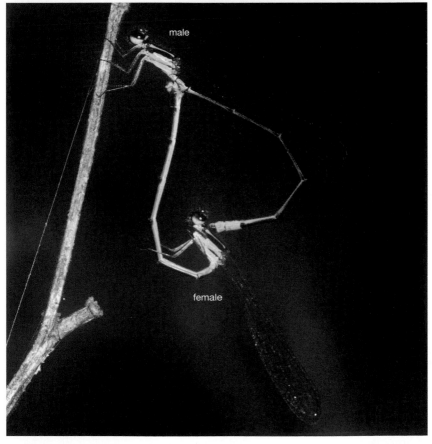

male

female

TL 31 mm; HW 17 mm. **Identification:** Sometimes occurs with Costa Rican Wedgetail (p. 125) in Caribbean lowlands, but Pacific more likely to be at streams than Costa Rican. In Pacific males, "wedge" at tip of abdomen higher than in Costa Rican. Males always blue, unlike Costa Rican, which often has green on thorax, especially antehumeral stripes. Paired pale spots on prothorax better developed in this species than in Costa Rican. Females very similar to female Costa Rican, but note tip of S1 entirely pale in both female and male Pacific, black above in Costa Rican. Females can also be distinguished by humeral stripe, which narrows and comes to point at upper end; remains broad in Costa Rican. **Habitat and Behavior:** Spring seeps, small streams, and marshy ponds; prefers sunny areas. Two males seen flying at each other, head to head, and ascending to 3 meters above water. **Range:** Both slopes and Central Valley, to 1400 m. Honduras to Ecuador and Venezuela.

genus *Acanthagrion* (p. 124)
male appendages in lateral view

Narrow-tipped Wedgetail	**Costa Rican Wedgetail**	**Pacific Wedgetail**
Acanthagrion inexpectum	*Acanthagrion speculum*	*Acanthagrion trilobatum*
(p. 124)	(p. 125)	

PEARLWINGS genus *Anisagrion*

Two species. These small damselflies of highland marshes have black-and-green eyes. Their short stigmas are white-rimmed and glisten like a pearl in the sun. The long, forked structure at the tip of S10 and long paraprocts produce a prickly look at the tip of the male abdomen. Both sexes go through dramatic color changes as they mature, going from orange to black and blue.

Middle American Pearlwing *Anisagrion allopterum*

TL 34 mm; HW 17 mm. **Identification:** Mature individuals have a mostly blue thorax and all black abdomen, unlike any other pond damsels. In males, black humeral stripe varies in length and width, may be narrow and incomplete. Mature females recognized by whitish glaze on thorax and pruinose S8. Orange immatures quickly develop blackish coloration on posterior abdomen; as thorax and head markings turn blue, entire abdomen becomes black. Orange immatures can be confused with immature female forktails with orange thorax, but forktails have prominent black median stripe. Orange immature Red-tipped Swampdamsel (p. 142) even more slender and develops green stripes on thorax. **Habitat and Behavior:** Spring seeps, roadside ditches, and marshy ponds. Immatures are common in grassy marshes together with mature individuals. This is unusual, as immature odonates generally stay away from breeding habitats. **Range:** Both slopes, 500–1600 m. Honduras to Costa Rica.

male

female

immature male

immature female

TL 37 mm; HW 19 mm. **Identification:** Males, with extensive blue on abdomen, more likely to be confused with dancers or bluets, but note "prickly" abdomen tip and short, white-rimmed stigmas. With much blue on sides of basal abdominal segments as well as S8-9, not quite like any other pond damsel. Like Middle American Pearlwing (p. 127), males can have distinct black humeral stripe or not; this much variation is unusual among pond damsels. Mature females, with mostly bluish thorax and mostly dark abdomen, very similar to female Middle American and difficult to distinguish, even in the hand. Immature stages have not been documented; they may or may not be similar to Middle American. So far, Kennedy's and Middle American have not been found together. **Habitat and Behavior:** Breeds in open sedge and grass marshes, and sometimes near small streams. **Range:** Southern Pacific slope, in San Vito region, 1000–1500 m. Southern Costa Rica and western Panama.

male

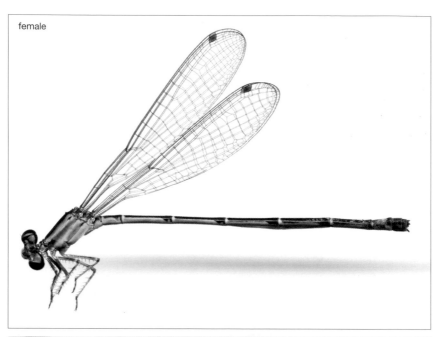
female

immature female

FIRETAILS genus *Telebasis*

Ten species. Firetails are so named because males of most species have a red abdomen, including all species in Costa Rica except for Golden Firetail. The members of this genus vary greatly in size, Golden and Montane distinctly larger than the other Costa Rican species, while Striped is one of the smallest damselflies in the region. Most species are associated with mats of floating vegetation.

Desert Firetail *Telebasis salva*

TL 26 mm; HW 14 mm. **Identification:** This and Striped the two smallest firetails in CR. In males, median black thoracic stripe divided by red carina and has hooklike projection extending from either side at upper end of stripe and small black dots and dashes lower on sides. Female brown, sometimes more reddish, but with same pattern on thorax. **Habitat and Behavior:** River flood pools, quiet pools in streams, and artificial ponds. Has greater habitat tolerance than other firetails; can be very common at floating mats of duckweed (*Lemna*) and water fern (*Azolla*). **Range:** Pacific slope, to 1000 m; also lowlands of southern Caribbean slope, near Panama border. Southwestern US to Venezuela.

male

female

TL 33 mm; HW 14 mm. **Identification:** Male Marsh and Hyacinth very similar, Marsh slightly larger. Both have pale antehumeral stripe so narrow that thorax looks mostly black above; both have red labrum and eyes and entirely red abdomen. Appendages provide clues to distinguishing the two; in Marsh, cerci about 70% length of paraprocts; in Hyacinth, cerci closer to 90% length of paraprocts and less curved downward. Female Marsh distinguished from female Hyacinth only on close inspection; in Marsh, ovipositor does not quite reach tip of abdomen, while it reaches tip in Hyacinth. Also note that Marsh female has a pair of very tiny projections that stick up at rear edge of prothorax (magnification necessary to see this). One clue to distinguish the two is habitat. Marsh is usually in emergent plants on pond margins, Hyacinth usually found associated with floating aquatic vegetation. **Habitat and Behavior:** Temporary pools and marshy wetlands in open or forested landscape. Often found in side channels and flood pools of streams with bordering grass and sedges. **Range:** Pacific slope, to 1200 m; Caribbean slope, to 800 m. Texas to Panama.

male

female

TL 30 mm; HW 15 mm. **Identification:** Males mostly red but thorax black above with one or two short, indistinct black stripes on either side. Compare with Marsh. **Habitat and Behavior:** Ponds in forest or open areas, usually entirely covered by aquatic plants or nearly so, especially water hyacinth (*Eichhornia*) and water lettuce (*Pistia*). **Range:** Pacific slope, to 500 m. Northern Mexico to Costa Rica.

male

female

TL 34 mm; HW 17 mm. **Identification:** Males very similar to male Red-and-black, the only other species that is red or orange with middle segments (S4–6) of abdomen dark and thorax bearing very fine black median line. In male Belize, eyes nearly black above and yellow below, while eyes mainly brick red in male Red-and-black. Note slightly curved male paraprocts, about 2x length of cerci. Female thorax has same fine black median line as male but shades from orange above to yellow below; abdomen ochre at base and increasingly darker toward blackish tip. **Habitat and Behavior:** Temporary pools, permanent ponds, and palm swamps, in and near forest. **Range:** Southern Pacific (Osa Peninsula) and Caribbean slopes, to 100 m. Belize to Costa Rica.

male

TL 35 mm; HW 18 mm. **Identification:** Males differ from Belize Firetail in having longer, straight paraprocts, about 3x length of cerci, and head with much more black above, except for red eyes. Also, S3 with much dark color, almost completely orange in Belize. Female a lot like female Belize but, as in male, more black on head. Red-and-black known only from Caribbean lowlands, but it appears to be more common than Belize, which occurs on both slopes. **Habitat and Behavior:** Ponds and palm swamps, in and near forest. Males and pairs increase in numbers at water in the afternoon, when they have been seen perched over water, in tandem, and foraging up in trees. **Range:** Caribbean slope, from Boca Tapada to Puerto Viejo area, to 100 m. **Costa Rican endemic.**

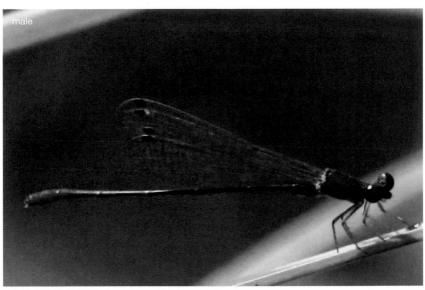

male

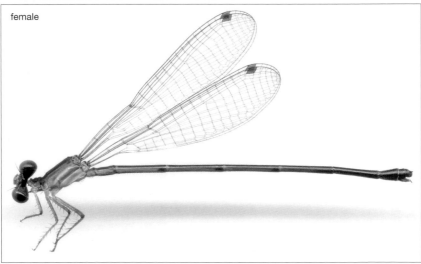

female

TL 30 mm; HW 16 mm. **Identification:** Males almost entirely red, with fine red markings on black head and wide, black median stripe on thorax divided by red carina. Thorax red to orange above, becoming paler below. Labrum and lower face dull green. Eyes dark red to black above and greenish below; only male Green-eyed (p. 140) and Montane (p. 139) have eyes similar to this species. In male, large, bulbous cerci and nearly invisible paraprocts are diagnostic at close range. Female similar in coloration to male, with red replaced by brown, and sides of thorax varying to greenish; split median thoracic stripe characteristic. Note superficially similar mature heterochromatic Rambur's Forktail (p. 118), with solid black thoracic stripe. Also called Coralline Firetail. **Habitat and Behavior:** Open ponds with emergent vegetation at shore. **Range:** Southern Pacific slope, to 1000 m. Costa Rica to Colombia and Brazil, West Indies.

male

female

TL 24 mm; HW 12 mm. **Identification:** A very small firetail, one of the smallest damselflies in Costa Rica. Males distinguished by combination of red abdomen and yellowish to greenish thorax with striped look not found in other firetails. Stripes formed from wide black median stripe split by pale carina, narrow antehumeral stripe, and wide black or brown humeral stripe narrowed behind to a point. Head black, with blue stripe across front; eyes black above and greenish below. Female a dull version of male, less vividly striped because antehumeral stripe always brown. Abdomen black above, light green to brown below; eyes brown above, yellowish below. Both sexes have distinctive narrow pale line along inner border of each eye. Female most easily mistaken for female forktail or bluet, but note distinctive thoracic pattern with pale median carina. **Habitat and Behavior:** Ponds with abundant floating vegetation, usually in forested areas. Both sexes often up in shrubs at forest edge. Rarely at marshy stream margins. **Range:** Both slopes, to 600 m. Central Mexico to Brazil.

male

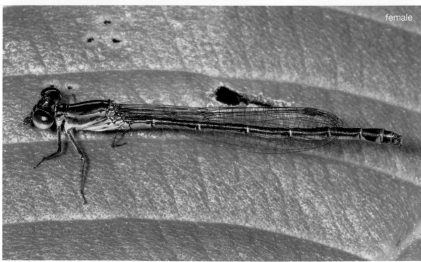

female

TL 38 mm; HW 21 mm. **Identification:** Both sexes unmistakable: mostly golden-orange, with vivid black markings on head and thorax. Females slightly duller than males. Possible confusion with largely orange immature Middle American Pearlwing (p. 127), but pearlwing lacks black humeral stripes of Golden. Orange immature swampdamsels smaller and more slender. **Habitat and Behavior:** Grassy areas around swampy ponds. **Range:** Southern Pacific slope, to 1500 m; so far, found only on Osa Peninsula and in San Vito region. Costa Rica and Panama.

male

TL 43 mm; HW 20 mm. **Identification:** A large firetail. Males and females have red abdomen but mostly greenish thorax; also note blue-green face and green on lower half of eyes. Wings usually tinted amber. Distinctly larger and at higher elevation than somewhat similar Green-eyed (p. 140), which has all green eyes and dark smudges on mid-abdomen. **Habitat and Behavior:** Temporary and permanent marshy ponds, marshy spring seeps, and pools in root holes of fallen trees. Males perch higher than other firetails, above dense vegetation. **Range:** Pacific slope, 1450–1600 m; Caribbean slope, 700–1600 m. Costa Rica to Ecuador.

male

female

TL 34 mm; HW 16 mm. **Identification:** Males distinctive, with blue face, green to blue-green eyes and thorax, and red abdomen. In some males, eyes mostly bright blue. Thorax has moderately wide black median stripe; very often, especially in younger individuals, central part of stripe paler, leaving two dark lines that define outer edges. Abdomen with variable amount of dusky wash, typically darkest on S6-7 and S10. Note possible confusion with some color variants of Red-tipped Swampdamsel (p. 142). Female brownish; thorax plain, with fine black lines on either side circumscribing an area equal to width of male's median stripe. Abdomen increasingly dark toward rear. In mature females, green eyes are distinctive; mature females of Belize and Red-and-black have reddish eyes. Brown females with brown eyes common, presumably less mature; not known if they mate in that stage. **Habitat and Behavior:** Marshes of all kinds, typically with dense grass. **Range:** Northern Pacific slope south to Parrita, to 300 m. Western Mexico to Venezuela.

male

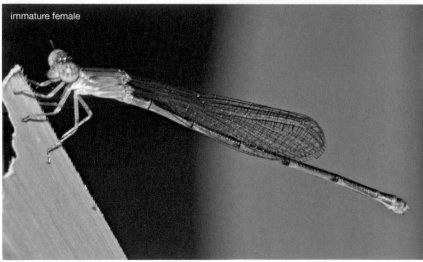

immature female

genus *Telebasis* (p. 131)
male appendages in lateral view

Belize Firetail
Telebasis boomsmae (p. 134)

Coral Firetail
Telebasis corallina (p. 136)

Marsh Firetail
Telebasis digiticollis (p. 132)

Striped Firetail
Telebasis filiola (p. 137)

Montane Firetail
Telebasis garleppi (p. 139)

Green-eyed Firetail
Telebasis isthmica

Hyacinth Firetail
Telebasis levis (p. 133)

Red-and-black Firetail
Telebasis rojinegra (p. 135)

Desert Firetail
Telebasis salva (p. 131)

no specimen available

Golden Firetail
Telebasis aurea (p.138)

SWAMPDAMSELS genus *Leptobasis*

Three species. These small, slender damselflies are common at seasonal wetlands. Two of the species show extensive orange coloration when immature, reminiscent of the conspicuously larger spinynecks and vaguely similar to threadtails with orange on the thorax or abdomen. They could also be confused with firetails, most of which are a bit shorter and more robust. Swampdamsels have prominent postocular spots, absent in firetails and spinynecks. Note the very short stigmas, as in pearlwings.

<h3 style="background:black;color:white;">Red-tipped Swampdamsel Leptobasis vacillans</h3>

TL 30 mm; HW 14 mm. **Identification:** Mature males and some females unmistakable, with black-striped bright green thorax and red-orange abdomen tip. Superficially similar in color pattern to Green-eyed Firetail (p. 140), with greenish thorax and black and orange abdomen, but that species is somewhat stockier, has entirely green to blue-green eyes, lacks postocular spots, has less defined humeral stripe, and shows less black on abdomen. Some female Red-tipped, presumably mature, similar in coloration to mature female Green-eyed Firetail, but lack black on head and thorax. This swampdamsel shows extreme variation, the reasons for which defy easy understanding but are probably related to wet and dry seasons. Individuals that lack black on head and thorax may be present only during dry season, then their offspring for one or two generations during wet season have black markings, then last eggs laid during wet season produce new generation without black. Immatures of both sexes entirely orange when very young; S4–7 or S5–7 quickly becoming blackish above, then thorax turns green and begins to develop dark stripes. **Habitat and Behavior:** Ponds and marshes in open areas or sunlit patches in forest. Both sexes and all age stages found in tall grasses at forest edge, sometimes in great numbers. Males display at each other with lowered abdomen tip, even bent at right angle. Surprisingly for such an abundant species, mating very rarely seen. Apparently adults present through dry season, mature females appearing at seasonal wetlands just before rains to lay eggs. Very rapid larval development (about three weeks from egg to adult) as soon as basins contain water, perhaps the quickest of all odonates. **Range:** Both slopes, to 900 m. Texas to Peru and French Guiana, West Indies.

male

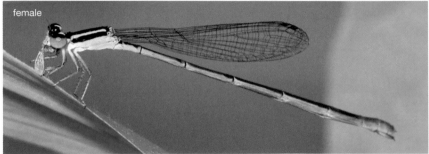

female

subadult male

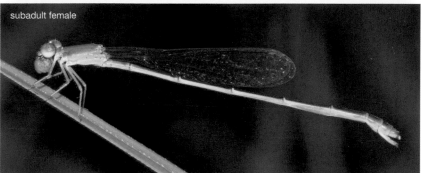

subadult female

immature male

immature female

143

TL 33 mm; HW 15 mm. **Identification:** Mature males distinguished by orange thorax with black median stripe and orange abdomen tip, a combination not found in male Red-tipped (p. 142). Postocular spots narrow and greenish in male Guanacaste, broad blue triangles in male Red-tipped. Male cerci longer than paraprocts in this species, shorter in Red-tipped, but small in both so not easy to see. Mature female thorax dark above and green below, without obvious markings; also note slender form. Orange immatures of both this and Red-tipped very similar, undergo similar changes with aging, and have to be distinguished by structure, especially at the abdomen tip. Black develops quickly on orange immatures of Guanacaste, then they become more like mature males, with orange markings low on S8-9 and entirely orange S10. Some individuals, however, keep much orange into maturity. Female Guanacaste has ovipositor shorter and more curved than in Red-tipped, but terminal abdominal segments much more expanded, giving abdomen tip a more bulbous look. **Habitat and Behavior:** Ponds and flooded river margins in forested landscape. Emerge in great numbers at beginning of rains, foraging among tall grasses under canopy. Found more often in forest than Red-tipped. Mating observed rarely, oviposition not at all. **Range:** Lowlands of Northern Pacific slope, only in Guanacaste Province. Western Mexico to Costa Rica.

male

female

TL 34 mm; HW 15 mm. **Identification:** Mature individuals show coloration quite distinct from that of other two swampdamsels. In males, thorax brightly striped black and green, abdomen black with yellow-orange tip (S7–10); females pale overall, with faint brown stripes on green thorax and blue abdomen tip (S8–10). Ovipositor straight, extending to tips of cerci. Immatures not known but perhaps orange as in other two species. Note: Florida males have reddish abdomen tip, as in the image here. No photos are available from the Central American range, where the abdomen tip is paler. Also, photograph of female does not show typical blue tip on abdomen. **Habitat and Behavior:** Swamp pools in forest. Both sexes found in grass along with more common Red-tipped; also forages among fallen tree branches at forest edge. Mating has not been observed; oviposition takes place in several soft-stemmed plant species, from near water to well above it. **Range:** Rare, known only from La Selva Biological Station but not seen there for many decades, possibly extinct in Costa Rica. Yucatan Peninsula, Mexico, to Costa Rica; also southern Florida.

male

female

SPINYNECKS genus *Metaleptobasis*

Two species. Both species are long and slender with bright orange, green, and black eyes, an orange thorax with a black median stripe, and a black abdomen. They typically perch with abdomen about 45° below horizontal. To distinguish the two, you must compare morphology, up close or in hand. They are long enough that individuals hanging from a forest leaf might be mistaken for a small helicopter damsel, but no other Costa Rican pond damsels are exactly like spinynecks. Their slenderness makes them something like *Protoneura* threadtails (pp. 151-153), but in the threadtails the thorax is orange above with extensive black on the sides. Males are distinguished from all other damselflies by the paired spines (thus the common name), also called horns, at the front of the pterothorax, but these can only be seen up close. There is no information on mating or oviposition of these species.

Guatemalan Spinyneck *Metaleptobasis bovilla*

TL 44 mm; HW 23 mm. **Identification:** Males distinguished from male Panamanian Spinyneck by thoracic spines shorter and thicker and projecting to greater degree over prothorax; note spines curved outward at tips. In Panamanian, spines long, very slender, and pointed, extending straight up above thorax. Male Guatemalan has very short cerci, one third length of paraprocts (about half length in Panamanian); paraprocts angled down just past mid-length in Guatemalan, smoothly curved downward in Panamanian. Females more similar to each other than males, but female Guatemalan has ovipositor valves extending distinctly beyond abdomen tip, while those of Panamanian just reach it. Female Guatemalan lacks thoracic spines, but, oddly, female Panamanian can either have them or not. In both sexes, Guatemalan has black labrum, Panamanian orange; close view of face essential to see this. The two species have been found to overlap only in the Caribbean drainage of Lake Nicaragua. **Habitat and Behavior:** Margins of swamps, shaded ditches, and slow streams. Both sexes perch on shaded leaves or vines up to head height. **Range:** Caribbean slope, to 300 m. Guatemala to Costa Rica.

TL 43 mm; HW 21 mm. **Identification:** Compare with Guatemalan Spinyneck. **Habitat and Behavior:** Males perch in shade with abdomen hanging down, either low in tall grass near water or high in shrubs in forest understory. One male was seen 1 m above the water in a swamp. **Range:** Three isolated regions known so far: Pacific slope on Osa Peninsula and Caribbean slope in drainage of Lake Nicaragua and near Panama border, to 300 m. Costa Rica to Venezuela.

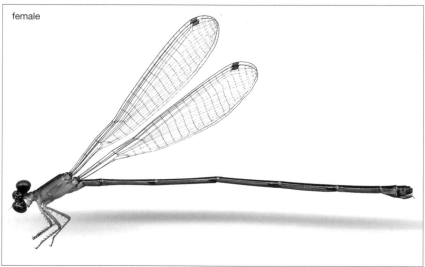

TL 26 mm; HW 13 mm. **Identification:** A tiny, slender damselfly. In both sexes, thorax is mostly blue and abdomen has blue at tip; also note bold black bar across top of head. Dusky spot on side of thorax instead of pale antehumeral and black humeral stripes distinguishes this species from most other small pond damsels. Immature similarly patterned but all pale areas tan. Other tiny pond damsels include some forktails and firetails, none closely resembling this species. **Habitat and Behavior:** Marshy ponds in both open and forested areas. Often difficult to see within dense grass or other vegetation. Mating rarely seen; oviposition in tandem or not, substrates including floating twigs and clumps of soil. **Range:** Pacific slope, to 1100 m; Caribbean slope, to 300 m. Florida, West Indies, and Mexico to Brazil.

male

female

THREADTAILS   genera *Neoneura*, *Protoneura*, and *Psaironeura*

Seven species. These small damselflies have narrow wings, perhaps adapted for prolonged hovering, at which they are adept. *Neoneura* are robust like typical pond damsels, but *Protoneura* and *Psaironeura* are very slender, thus the common name. All of them were formerly placed in a separate family, Protoneuridae. Several species of *Protoneura* are known to form leks, with groups of males hovering among the trees near their breeding habitat and females finding them there. Note black head with red eyes (yellow beneath) in all three *Protoneura* species. Female *Protoneura* have a distinctive shape, with very slender abdomen becoming wider in posterior half; and they oviposit in a distinctive way, with female abdomen bent sharply at two places, S6 nestled between the wings, which hold it in place. Copulation is rarely seen in *Protoneura* and *Psaironeura*; perhaps it takes place away from water or high in trees.

Amelia's Threadtail   *Neoneura amelia*

TL 32 mm, HW 15 mm. **Identification:** Males only damselflies with red-orange head, thorax and abdomen base; rest of abdomen black. Female Amelia's and female Esther's (p. 150) are the only brown damselflies that have no dark markings except narrow and faint dusky streaks on thorax and even less distinct markings on abdomen. Females of these two species not distinguishable in the field, but each usually seen in tandem with male. In female Amelia's a touch of red sometimes present at rear of head, so far not seen in female Esther's; also note that some Esther's have richer orange-brown color on abdomen than Amelia's. **Habitat and Behavior:** Occurs in forested and semi-open areas, in pools in small rivers and streams. Males very often hover a few centimeters above water but also perch in shrubs well above water. Tandem pairs oviposit in floating wood chips and plant debris, sometimes together in numbers. **Range:** Both slopes, to 600 m. Texas to Costa Rica.

male

female

TL 32 mm; HW 14 mm. **Identification:** Males only damselflies with irregularly striped reddish and black thorax and red abdomen. Crimson Threadtail much more slender and more brightly colored. Male firetails can look similar, but all have red, green, or partially green eyes and thoracic patterns that differ from that of Esther's. Compare female with female Amelia's (p. 149). **Habitat and Behavior:** Pools in small streams running through forest. More sluggish than Amelia's, hovering less often over water. Tandem pairs oviposit in floating plant material and rootlets along bank. **Range:** Pacific slope, to 400 m. Nicaragua to Venezuela.

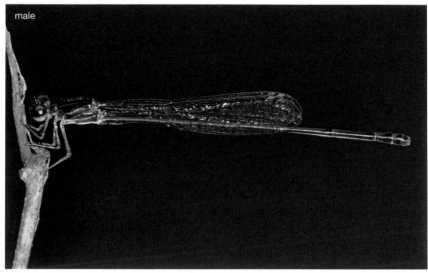

male

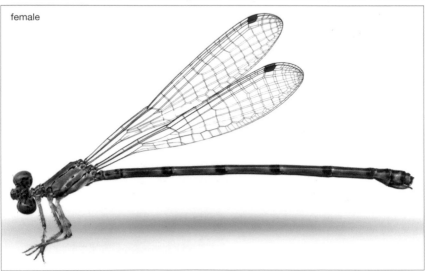

female

TL 36 mm; HW 17 mm. **Identification:** Males are the only threadtails that are entirely brilliant red. Only other red damselflies are male firetails, but they are shorter and much less slender. Females of the three species of *Protoneura* are quite similar, but this one has more black on thorax, with only a narrow pale antehumeral line, while the other two have more orange or yellow in same area. **Habitat and Behavior:** Pools in small streams running through forest, where males hover at water surface and up to 2 m above; they also hang on leaf tips of overhanging shrubs. Six males seen in what was probably a lek, hovering 2–5 m above muddy ground near stream and shrubs, all facing same way and slowly ascending and descending. At another site, individual males interacted, one dropping onto the other while facing same way, rubbing abdomens as though rubbing sticks to start a fire. Tandem pairs often seen flying over water or ovipositing. **Range:** Both slopes, to 400 m. Guatemala to Venezuela and Ecuador.

TL 36 mm; HW 17 mm. **Identification:** Males distinctive, with bright orange thorax and mostly black abdomen. In females, note typical abdomen shape, wider toward rear, and orange and black stripes on thorax. Compare with female Sulfury for subtle distinctions. **Habitat and Behavior:** Wide pools in small forest streams. Behavior similar to that of Crimson Threadtail. **Range:** Caribbean slope, to 400 m. Eastern Mexico to Panama.

TL 34 mm; HW 16 mm. **Identification:** Males similar to male Golden-orange—both have bright thorax and black abdomen—but thorax yellow instead of orange, and black median stripe narrower. S3 mostly black (mostly orange in Golden-orange). Golden-orange has orange spot on front of head, lacking in Sulfury. Females much like female Golden-orange but generally with more black on thorax. Both female Sulfury and female Golden-orange show differences in patterning on the thorax from female Crimson (p. 51), in which it is mostly black in front, with fine pale antehumeral lines. **Habitat and Behavior:** Pools in small streams running through forest; also found at tiny trickles in shade and open water of swamps. Behavior very likely similar to that of Crimson Threadtail, although one female seen ovipositing by herself. **Range:** Southern Pacific slope north to Carara NP and Caribbean slope, to 400 m. Nicaragua and Costa Rica.

male

female

TL 34 mm; HW 16 mm. **Identification:** Coloration of Wispy and Selva males unlike that of any other small damselfly: black overall, with reddish eyes and lower thorax, and pale pruinose abdomen tip. Eyes of male Wispy are not entirely red, as in male Selva. Females duller, somewhat like females of *Protoneura* threadtails (pp. 151–153) but lack stripes on thorax. Ovipositor valves just reach tips of cerci. **Habitat and Behavior:** Swampy ponds, spring seeps, and pools in small streams in forest. Perches in shade. Males hover for long periods over water, when pale spot at abdomen tip is all that can be seen from a distance. In display, two males will sometimes bob up and down within 10 cm of each other. **Range:** Both slopes, to 900 m. Nicaragua to Ecuador and Venezuela.

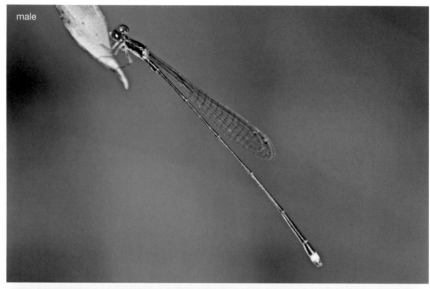

male

female

TL 34 mm; HW 16 mm. **Identification:** Males very similar to male Wispy, but note eyes entirely red; upper part of eyes somewhat darker in Wispy. Male cerci distinctive if seen well; ventral lobe narrow and indented above in Selva (like a horse's head in profile), wider and not indented in Wispy (like a cat's head). In immature male Selva, much of abdomen red, a trait not observed in immature Wispy. Selva immatures have been mistaken for Red-tipped Swampdamsels on first sight. In female Selva, ovipositor distinctly longer, projecting beyond tips of cerci; in female Wispy, ovipositor and cerci even. **Habitat and Behavior:** Ponds and slow moving, small streams, usually bordered by forest but sometimes by grass or open areas. Behavior like that of Wispy Threadtail, spending most of time perched in herbaceous vegetation but at times hovering over water within 0.3 m of surface. Males interact by hovering very close to one another, either parallel or face-to-face, with abdomen inclined below horizontal; they "bounce" rapidly up and down just a few centimeters, with red eyes and pruinose abdomen tips quite evident. **Range:** Caribbean slope, to 600 m. **Costa Rican endemic**.

male

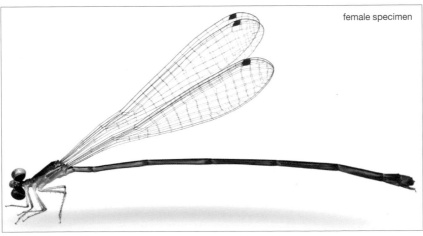

female specimen

155

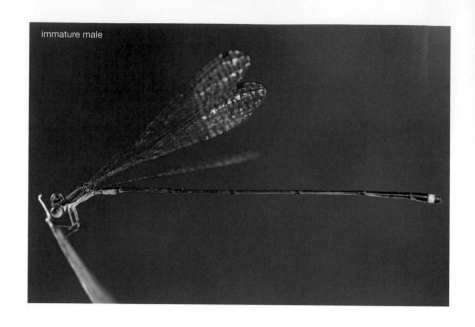

immature male

genus *Psaironeura* (p. 149)
selected male appendages in lateral view

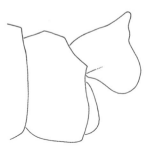

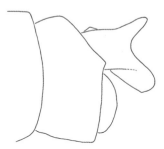

Wispy Threadtail
Psaironeura angeloi (p. 154)

Selva Threadtail
Psaironeura selvatica (p. 155)

HELICOPTER DAMSELS genera *Mecistogaster*, *Pseudostigma*, and *Megaloprepus*

Five species. These are the largest damselflies. Their colored wing tips seem to whirl about each other rather than beat up and down, thus the common name. When flying slowly, the wing motion is easy to follow due to the colored tips. The forewings scull through the air above the hindwings, with the abdomen not much below the horizontal. *Mecistogaster* species have rather narrow wings, those of *Megaloprepus* and *Pseudostigma* much broader. All breed above ground in phytotelmata: bromeliads, tree holes, and crevices in standing or fallen trees that hold water. Thus, they are found only in forests with enough rainfall in the wet season to maintain these larval habitats. Lemon-tipped is the only species found in drier forests; all species, however, seem to be capable of surviving the six-month dry period between wet seasons as adults. Helicopter damsels are specialized foragers, plucking spiders from webs while in flight; a crystalline wax covering on their wings apparently helps protect them from adhesive spider threads. They vary in size more than other damselflies, because larval growth is more limited in smaller spaces or growth rate is speeded up as their microhabitats dry out. Therefore a range of measurements is stated rather than an average. Wing-tip markings are important for identification, but they vary by both sex and age. Markings that look like a stigma are a collection of pigmented cells called a pseudostigma.

Bromeliad Helicopter *Mecistogaster modesta*

TL 74–83 mm; HW 38–45 mm. **Identification:** The smallest helicopter, yet still larger than any other pond damsel. In male forewing, pseudostigma 6 cells long in first row and 3 cells long in second; hindwing has only a front row. In immature males, wing tips white; as individuals mature, white turns to yellow, then orange-brown (dark brown in some individuals). In immature females, wing tip hazy white (because of white veins) with snow-white pseudostigmas; when older, wing tip becomes clear, with reddish pseudostigmas. Mature Bromeliad distinguished from Long-tailed (p. 158) and Lemon-tipped (p. 159) by reddish pseudostigmas and completely black underside of thorax. Also, male cerci more slender than in other *Mecistogaster*. **Habitat and Behavior:** Wet forest, individuals usually foraging along edges of clearings exposed to sunlight. Larvae live in tank bromeliads; this and Canopy Dragonlet the only CR odonates known to use them. **Range:** Both slopes, to 1200 m. Eastern Mexico to Ecuador and Venezuela.

male

female

TL 109–137 mm; HW 47–64 mm. **Identification:** Males stand out as the most elongate of the clear-winged helicopters, with abdomen more than twice wing length; females are quite a bit shorter than males and within the size range of the other species. Male wing tips milky white after emergence, becoming clear within a week except for small (about 6 cells in front row; usually 2 cells in second row) pseudostigma that becomes black in mature adults. Female wing tips and pseudostigma white to pale yellow when immature, later developing markings like that of male. Underside of thorax with black stripe on midline. Even smallest males much larger than male Bromeliad (p. 157); females also larger than female Bromeliad. Also note that mature Bromeliad has colored pseudostigmas, not black. Distinguished from Lemon-tipped by smaller markings at wing tip and more vivid black and green striping on thorax. Distinguished from similarly colored Broad-winged (p. 160) by narrower wings and smaller pseudostigmas. **Habitat and Behavior:** Wet forest, where individuals can be encountered anywhere from near ground to well up in canopy, fluttering and hovering with wing tips blurring around them. **Range:** Caribbean slope, to 300 m. Nicaragua to Bolivia and Brazil.

male

female

TL 73–97 mm; HW 41–55 mm. **Identification:** In male, wing tips bright yellow, becoming black on underside in breeding season. In female, wing tips also bright yellow (even when breeding), bordered by narrow dark line; with age develops brown wash on outer part of wings. Only species of helicopter lacking dark stripe low on side of thorax. Among helicopters, only this and Long-tailed show black stripe on midline under thorax. Note yellow wing tips of larger and much rarer Broad-winged (p. 160), which has more vivid thoracic striping. Body of Lemon-tipped brown overall, drabber than other helicopters. Up close, note that male is only one of the four clear-winged helicopters with pale tibiae. **Habitat and Behavior:** Wet forest in the southern Pacific lowlands but also in Guanacaste dry forest, where considerably drier; adults there presumably survive long periods between wet seasons. Breeds in tree holes, which must maintain enough water to support complete larval development. Much foraging just above ground through herbaceous vegetation, lower than other helicopter damsels, but also ascending into trees. **Range:** Pacific slope, to 1400 m, extending to Caribbean slope in far northern Guanacaste (Volcan Orosi). Northern Mexico to Argentina and Brazil.

male

female

TL 126–145 mm; HW 62–73 mm. **Identification:** Very large, broad-winged damselfly; distinguished from Blue-winged by mostly clear wings with yellowish tips. In mature male, wing tips black only on front edge of wing, covering slightly more than one cell row; front edge pale in immature males. Female wing tips pale yellowish, with dark spot equivalent to male's larger black area; with age becomes quite dark proximal to yellow tip. Broader wings than similarly long-bodied Long-tailed (p. 158) and Lemon-tipped (p. 159) Helicopters, but Lemon-tipped above the observer and thus with wings foreshortened can look very much like it. Broad-winged has narrow black stripe low on thorax that is absent in Lemon-tipped. **Habitat and Behavior:** Forages in a manner similar to *Mecistogaster* and *Megaloprepus*; seen fluttering slowly through sunny clearings in primary forest. Breeding sites unknown, perhaps very limited, which would explain it being uncommon and scarcely ever seen in CR. **Range:** Mostly single records from five localities, two on central Pacific slope (Carara NP) and three on Caribbean slope, to 800 m. Eastern Mexico to Panama.

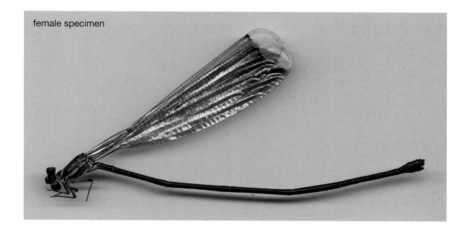

female specimen

TL 75–115 mm; HW 61–85 mm. **Identification:** A very large and distinctive damselfly, with the longest wings of any odonate, but adults quite variable in size. In males, note blue-black wing band and white inner band bordering it. Males from Pacific side up to Monteverde look like females, with no white band. Females average considerably smaller than males, with relatively shorter abdomen. **Habitat and Behavior:** Primary forest, breeding in tree holes. Flight slow and fluttery but surprisingly elusive when disturbed. Males set up territories around tree holes, defending against other males and mating with females that come into their territory. Females oviposit at edge of water, not in it. Adults very long-lived; one marked individual lived for six months, a record among odonates. **Range:** Both slopes, to 1500 m. Eastern Mexico to Bolivia and Guyana. (Recent studies have shown that there are actually two species of *Megaloprepus* in Costa Rica, one consisting of individuals on the Pacific slope, the other of those on the Caribbean slope. A formal description of this taxonomic change has yet to be published.)

male Caribbean slope

male Pacific slope

female

Darners (family Aeshnidae)

Twenty-seven species. At any given spot, darners are usually the largest dragonflies to be found, with big eyes and long abdomens. They fly incessantly, including above the forest canopy, while foraging or looking for a mate. Like other flier dragonflies, they perch by hanging from a twig or vine, with the abdomen almost vertical. Most species breed in ponds, a smaller number in streams. Some genera (*Gynacantha, Neuraeschna, Triacanthagyna*) roost within the forest and forage at twilight (sometimes also at dawn), coming out to feed at the forest edge. Many individuals can make up these flights, but the dusk-fliers are not often seen when at their breeding sites, in rainpools in the forest.

The big pilot darners are the darners most often seen flying around clearings during the day; they vary from green and black to mostly red. At higher-elevation marshes, the green-striped Malachite Darner and blue-striped Black-tailed Darner can be common. The huge megadarner is unmistakable by size alone. Most members of this family have a prominent marking on top of the frons, often T-shaped, called a T-spot; it is sufficiently variable in some genera for use as a distinguishing character. In most species, the wings become suffused with brown as they age; brown-winged individuals may live for several months, probably longer, which is quite old for an odonate. A few species migrate seasonally within Costa Rica.

TWO-SPINED DARNERS   genus *Gynacantha*

Nine species. Members of this genus vary in size and color. In older individuals, the wings become dark brown, but from above note colored sclerites at the wing bases. All species have pale legs except for Golden-tipped (black) and Chartreuse (orange with black tips). Females of all species have a two-pronged process under the abdomen tip that perhaps serves to position the ovipositor when laying eggs or excavating the mud in which some species oviposit.

Two-spined have larger eyes than most other darners, which help them see in dim light, and most species are crepuscular, actively feeding for a half hour at dusk and at dawn, when forest clearings and roadsides suddenly become full of them. They often roost in and around buildings and at night regularly come to lights or even enter open windows. Males of two species, the brightly marked Gold-tipped and Chartreuse, fly over breeding territories in the midday sun, but the others remain within forest, where they roost during the day. Two-spined are rarely seen mating. Most oviposit in fallen logs, more rarely in mud at the water's edge or in completely dry basins in swamps that fill during the wet season. A few breed opportunistically in tree holes. The nine species in Costa Rica fall into three size groups: large (Pale-banded and Dark-saddled), medium (Auricled, Bar-sided, Twilight, and Gold-tipped), and small (Yellow-legged, Chartreuse, and Little Brown). Species of the last group are as small as three-spined darners, *Triacanthagyna*.

TL 74 mm; HW 50 mm. **Identification:** Brown darner with no wing markings; greenish eyes and green-tinged thorax when mature; in older individuals, wings become more suffused with brown. Note two or three small black spots on each side of thorax. Green markings between wings and at abdomen base and less constricted abdomen distinguish both sexes from Bar-sided (p. 166) and Auricled (p. 167), both with blue markings. Individuals much like Twilight Darner but conspicuously larger are known from both slopes at higher elevation (1200–1400 m). They differ in having a faint dark bar on lower edge of each side of thorax and a dark stripe along anterior margin of each wing; may represent an undescribed species. **Habitat and Behavior:** Breeds mainly in swamps and on the Pacific slope, also in small seasonal streams. Individuals roost in forest but fly out to forest edge and even open areas at dusk and dawn, often traveling substantial distances; foraging typically lasts for a half hour. Reproductive activity takes place only during day. Females oviposit in muddy soil in low areas in gallery forest that sometimes flood. Often found in and on buildings. Common in Monteverde, where it does not breed, during apparent seasonal migrations, with November the peak month. **Range:** Both slopes, to 1500 m. Southern US and West Indies to Bolivia and Brazil.

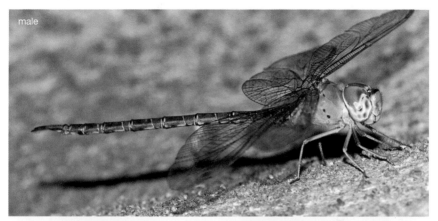

male

female

TL 70 mm; HW 45 mm. **Identification:** Brown darner similar in size and color to the more common Twilight Darner and with similar small black spots on each side of thorax, but in addition a dark horizontal bar below the spots. In both sexes of Bar-sided, base of abdomen more constricted than in Twilight. Mature males have blue markings between wings and on abdomen base; similar markings green in Twilight. Twilight never shows dark line extending out on each wing that many Bar-sided show. Female eyes duller than those of male and abdomen less constricted. See also Auricled Darner. **Habitat and Behavior:** Lives in forest and forages at dusk, but usually much less common than Twilight. Reported ovipositing in muddy donkey hoofprints. **Range:** Both slopes, to 1500 m; best known from Guanacaste lowlands but also found at Manuel Antonio NP, Monteverde, and La Selva Biological Station. Texas to Peru and Brazil.

male

female

TL 70 mm; HW 50 mm. **Identification:** Males brown overall, with green eyes and slightly greenish thorax; very similar to male Twilight (p. 165) and Bar-sided. Indeed male Auricled almost identical to male Bar-sided, both with same size, wasp waist, bits of blue at abdomen base, and epiproct that is less than half length of cerci; differs by lacking prominent dark markings on each side of thorax. When side view not possible, note that T-spot on frons smaller in Auricled than in Bar-sided, stem narrower, and crossbar at top less than half width of frons (more than half width in Bar-sided). Slightly smaller than Twilight, which has no blue on S3 and slightly less constricted abdomen base, as well as epiproct in male longer than half length of cerci. Female Auricled and female Twilight both with abdomen not constricted basally, but in female Auricled eyes are more green than brown, no black spots on side of thorax, and more blue at base of abdomen. **Habitat and Behavior:** A forest-based species like others of genus. In one observation, males patrolled ditch containing water at dusk, and at least one female oviposited there at the same time. **Range:** Caribbean slope, to 100 m; recorded from La Selva and Tortuguero. Belize to Bolivia and Brazil.

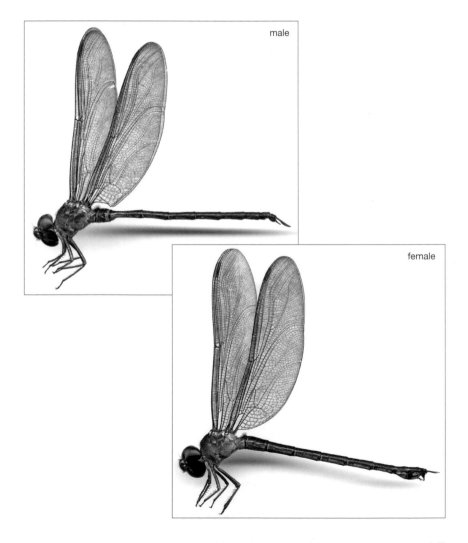

male

female

TL 57 mm; HW 35 mm. **Identification:** Very small and very plain, with abdomen not narrowed at base. Yellow-legged Darner, which is also small, shows bright markings. Little Brown most like Pale-green Darner (p. 178), equally drab and with pale legs, but distinctly smaller. Male cerci distinctive: well separated, narrow and curved, ending in wide lobes at tip. **Habitat and Behavior:** Presumably forest-based, but biology unknown and rarely seen, so far only in dusk flights. **Range:** Pacific slope, to 100 m; recorded only from Rio Tempisque area. Eastern Mexico to Brazil.

male specimen

TL 55 mm; HW 38 mm. **Identification:** Only brightly colored darner with yellow legs, but note that some dull species also have yellow legs. Shares green stripes on brown thorax with larger Chartreuse Darner (p. 170), but both are quite different from all other dusk-flying darners with stripes; latter have narrow brown stripes on green thorax. Male coloration somewhat similar to that of male Highland and Turquoise-tipped Darners but differs from them in shape, with much more slender and wasp-waisted abdomen. Like other female *Gynacantha*, females have spiny prong at abdomen tip (prominent in side view). **Habitat and Behavior:** Small streams and seeps within forest. Apparently quite rare, daily cycle not known. Seen hanging up during day in deep forest shade along streams. **Range:** Caribbean slope, to 300 m; also recorded from northern Guanacaste (Santa Cecilia) south to Horquetas, but presumably more widespread. Costa Rica to Ecuador.

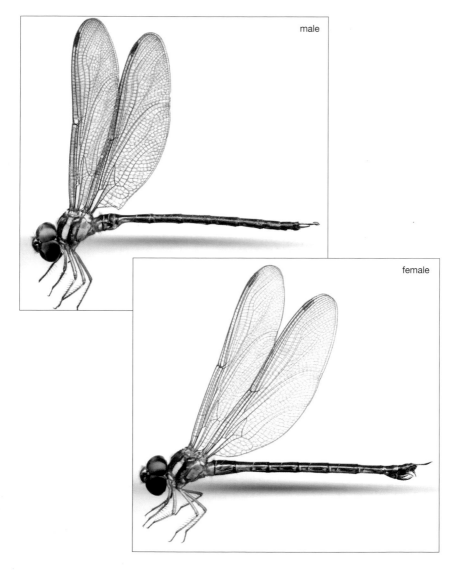

male

female

TL 61 mm; HW 43 mm. **Identification:** An unusual two-spined darner. Male Chartreuse and male Gold-tipped both active during day and brightly colored but with different pattern. In male Chartreuse, note bright green thoracic stripes; also seen in male Mountain Stream, Highland, and Turquoise-tipped Darners, abdomen distinctly more slender than in those species. Similar in shape to Yellow-legged Darner but much more brightly colored, with more vividly striped thorax and dark legs. **Habitat and Behavior:** Perhaps breeds in pools in seasonal forest streams, where both sexes have been found, including a male patrolling over such a pool. Female observed foraging in forest clearing at midday. Known so far from only four specimens. **Range:** Caribbean slope, 500–850 m. **Costa Rican endemic.**

male

female specimen

TL 71 mm; HW 48 mm. **Identification:** Only darner with thorax green and abdomen shading from blue or green through brown to orange to yellow at tip, in both sexes. Males with blue eyes, females green. **Habitat and Behavior:** Breeds in open ponds in forest. Males alternate between flying and hovering, moving low over open water and aquatic vegetation; they hang up low in shrubs over land or water. Usually hover briefly before hanging up. Most CR darners do not hover during sexual patrol flight. Females oviposit on soggy wood chips in shade or sun, in moist areas but not in standing water. **Range:** Southern Pacific and Caribbean slopes, to 700 m. Costa Rica to Peru and Venezuela.

male

female

171

TL 80 mm; HW 58 mm. **Identification:** Quite large and both sexes very similar to Dark-saddled Darner, which occurs in same areas and has similar habits, but black patch at hindwing base much smaller than in Dark-saddled. Dark-saddled also differs in having green thorax with large dark area below hindwings and no pale markings on most of abdomen; Pale-banded has plain thorax and pale bands across abdomen. Males with blue eyes and blue sclerites between wings, while male Dark-saddled have green eyes and usually green sclerites. Female Pale-banded and female Dark-saddled even harder to tell apart than males but note that female Dark-saddled has darker abdomen; also look for distinctive wing markings. In addition, tibiae reddish-brown in Pale-banded, pale in Dark-saddled, especially when young, and frons of Pale-banded with distinct T-spot, all black in Dark-saddled. **Habitat and Behavior:** Forested ponds and swamps. Males fly during daytime and hover in dense swampy areas, often among *Spathiphyllum*. Hang up suddenly, often quite low, with no hovering as in Gold-tipped. Sometimes fly high into trees when flushed. Females oviposit in mud and wet mossy logs above water in swampy areas, at any time of day. **Range:** Both slopes, to 1200 m. Guatemala to Argentina.

male

immature female

172

TL 80 mm; HW 55 mm. **Identification:** Large darner with prominent dark patch at base of hind-wing; compare with Pale-banded. Male eyes green, female eyes blue. Female abdomen base less constricted than in male. **Habitat and Behavior:** Forest dweller. Known to breed in tree holes, where larvae have been found, but also in shallow pools on forest floor that fill during wet season. Active during afternoon, males searching widely for breeding habitat and females, including along forest streams, where they may also breed in pools. During morning, males commonly perch on shrub stems in shade next to small forest pools. **Range:** Both slopes, to 800 m. Nicaragua to Bolivia and Brazil.

male

female

173

THREE-SPINED DARNERS genus *Triacanthagyna*

Four species. The members of this genus are less varied than the two-spined darners (*Gynacantha*). All are about the same size, with a green thorax, with or without brown stripes, and a relatively long, slender abdomen. Their wings darken with age and in some cases become heavily suffused with brown. Females of all species have a three-pronged process under the abdomen tip, compared with two in two-spined. All are crepuscular, with relatively large eyes, but they often fly higher and more erratically when foraging than two-spined and often appear earlier in the evening when it is lighter. They appear in and around human habitations, where they sometimes forage close to the ground at dusk. They hang up fairly low in the forest understory when inactive. Like two-spined darners, they often oviposit in dried-up basins in swamps that fill in the wet season.

Caribbean Darner *Triacanthagyna caribbea*

TL 63 mm; HW 41 mm. **Identification:** In both sexes, green thorax with brown stripes. Abdomen blackish to red-brown with light green markings. Some males with S10 distinctly pale, even yellowish; this not duplicated in Satyr and slightly smaller Ditzler's. Another color pattern distinctive of Caribbean is found on the legs: inner surface of first and second femora pale green, contrasting with entirely brown third femur, usually only visible in hand. Other two species show no such contrast between second and third femora, which are brown. See Ditzler's (p. 176) and Satyr (p. 177) for more details. **Habitat and Behavior:** Widespread and probably most common of brown-striped members of genus. Common in early dry season (November-December), when dragonflies migrate through Monteverde. **Range:** Both slopes, to 800 m; to 1450 m when migrating; appears to be more common on Pacific slope. Texas to Bolivia and Brazil, Lesser Antilles.

male

female

male

TL 59 mm; HW 36 mm. **Identification:** Essentially identical to Caribbean (p. 174) and Satyr, but distinctly smaller and apparently less common. Anal loop usually with 2 cells at front border and as few as 5 total cells in loop (cells in 2 rows); other species typically with 3 cells at border and total of 9–11 cells (cells in 3 rows); unfortunately, cell count is variable and not always diagnostic. Hamular processes of this species shorter than those of the other two, only visible by close examination in hand. Females of all three species very difficult to distinguish even in hand, though Ditzler's is slightly smaller than Satyr and, as in males, typically has fewer cells in anal loop. **Habitat and Behavior:** Presumably breeds in small ponds in forested areas, roosts in forest understory. **Range:** Both slopes, to 300 m. Southern Mexico to Bolivia and Brazil.

male

female

TL 63 mm; HW 41 mm. **Identification:** Very similar to Caribbean (p. 174) and Ditzler's; compare leg coloration in Caribbean and wing venation in Ditzler's. To confirm identification of male in hand, look for patch of tiny black denticles on lower margin of genital lobe on S2, not usually present in other two (male Caribbean may have a few denticles, usually smaller and paler). **Habitat and Behavior:** Temporary ponds in forest. Females oviposit in soggy wood of swampy areas, not always near water and often in dense tangled vegetation that makes flight difficult. **Range:** Both slopes, to 800 m. Belize to Peru and Brazil.

male

female

TL 58 mm; HW 38 mm. **Identification:** Only member of genus with pale legs and thorax without stripes; eyes blue or green, duller in females. Base of abdomen less constricted than in other three-spined darners, not at all in female. Superficially resembles other darners with all green thorax such as Blue-faced (p. 182), but pale legs and more slender abdomen distinctive. **Habitat and Behavior:** Breeds in forest ponds, forages throughout forests and well out from their edges. Seen mating in forest far from water. They become common in August-September in some forested stream canyons on the Pacific slope (600–800 m), perhaps seeking humid refuge, then apparently migrate east through Monteverde (1450 m) in large numbers well into the dry season. **Range:** Both slopes, to 800 m; to 1450 m during migration. Texas, Florida, and West Indies to Bolivia and Brazil.

male

female

TL 88 mm; HW 64 mm. **Identification:** In both sexes, large size and narrow yellow-green thoracic stripes distinctive; also note pair of sky-blue patches on front of thorax in male. Small dark patches at wing bases, and female shows brown tint on wings from nodus to stigma. Eyes mottled brownish green in front. Only other darner of similar size is Magnificent Megadarner (which lacks wing markings). Pale-banded and Dark-saddled Darners (pp. 172-173) are also large, inhabit forest interior, have dark basal wing markings, and fly at dusk, but they lack the thoracic markings of this species. **Habitat and Behavior:** Found in lowland rainforest, presumably breeding in ponds and swamps. Poorly known and apparently uncommon. Like other dusk-flying darners, sometimes hanging up near light in morning or trapped in building with open window. **Range:** Caribbean slope, to 700 m. Honduras to Ecuador.

179

TL 100 mm; HW 66 mm. **Identification:** Combination of huge size, unmarked wings, glowing green eyes, and overall blue-gray to green coloration is unmistakable. Note that Guanacaste populations have blue-gray thorax and abdomen, all others green. In older individuals, wings turn brown. Cerci crooked in lateral view, straight in other darners. **Habitat and Behavior:** Males often seen patrolling over clearings, perhaps foraging rather than searching for females. Also observed patrolling up and down a variety of waterways, including tiny trickles, flowing pebbly streams, and large, slow rivers. Foraging/searching beats may be hundreds of meters long with slow and steady flight no more than 2 meters above water or ground. Female collected with mud on abdomen hints that oviposition site may be mud rather than plants. Note that females have a two-pronged process under the abdomen tip as in *Gynacantha*. **Range:** Both slopes, to 100 m. Guatemala to Argentina.

female

PILOT DARNERS genus *Coryphaeschna*

Five species. These large green or red darners are often seen in feeding flight over clearings, sometimes in mixed-species groups; they hang up in nearby trees to feed or rest. Females have long cerci, longer than those of males, that are usually broken off with maturity. Females uninterested in mating fly with the abdomen tip bent down. Female eye colors change with age, sometimes dramatically. All species breed in still waters in both open and forested habitats; also seen over streams.

Blue-faced Darner *Coryphaeschna adnexa*

TL 64 mm; HW 40 mm. **Identification:** Smallest pilot darner, most similar to larger Mangrove Darner (p. 184). In male Blue-faced, eyes green with blue rims; in male Mangrove, eyes also green but note green face. In female Blue-faced, eyes green with more extensive blue on face; in female Mangrove, note blue eyes and green face. Both sexes of Blue-faced also have less green on sides of middle segments of abdomen than Mangrove. At close range, Mangrove shows dark vertical dash on either side high on front of thorax, lacking in Blue-faced, which may show dark marking low and closer to suture lines on either side. Note that Blue-faced has short epiproct, less than one-third length of cercus, more than half cercus length in Mangrove. Other pilot darners with green thorax have mostly brown abdomen with faintly discernible rings. Blue-faced could also be confused with Pale-green Darner (p. 178), which has smaller thorax and thus abdomen looks relatively longer and more slender. Pale-green also differs by its brown abdomen with no center stripe, greenish to brown face that lacks T-spot, and pale legs. Note that very young immature Blue-faced have prominent brown thoracic stripes that disappear quickly with age but could cause confusion with the striped species of *Triacanthagyna*. **Habitat and Behavior:** Breeds in ponds and swamps with floating vegetation such as water lettuce, water hyacinth, and water fern covering much of surface. Foraging individuals tend to fly lower than the larger pilot darners and stay closer to forest. **Range:** Both slopes, to 400 m. Texas, Florida, and West Indies to Argentina.

male

female

TL 73 mm; HW 49 mm. **Identification:** A large darner, with entirely green thorax and green-ringed black abdomen. Mature males with green eyes; mature females with blue eyes; immature females with green eyes and long cerci. Blue-faced (p. 182) is most similar but is conspicuously smaller and has slightly different eye colors and different abdominal markings; see that species. Note that green darners (*Anax*, p. 193) also have green thorax but have abdomen spotted (not ringed) and more expanded at base. **Habitat and Behavior:** Breeds in ponds with emergent vegetation, often at edge of forest. Usually forages high in open air. Also seen in freshwater mangrove swamp on Pacific coast, a common habitat in West Indies but little surveyed in CR. **Range:** Both slopes, to 1200 m. Southern Mexico, Florida, and West Indies to Bolivia and Paraguay.

male

female

TL 81 mm; HW 52 mm. **Identification:** Large red darner of southern Pacific slope, similar only to Fiery Darner (p. 186) but slightly larger and with reddish wing veins (dark in Fiery and all other pilot darners). Many mature Amazon Red show some green on thorax, not the case in Fiery. Males have reddish face lacking T-spot. Females have eyes and face greenish to blue, thorax and abdomen base green, rest of abdomen mostly brown. Female Amazon Red with reddish basal wing veins and black hind tibiae, Fiery with brown wing veins and brown tibiae, but tibiae do become dark in some individuals of Fiery. Female Amazon Red often show faintly indicated T-spot on face and dark dashes on front of thorax, never the case in Fiery. **Habitat and Behavior:** Males fly over swampy ponds; both sexes forage widely away from water. **Range:** Southern Pacific slope, to 100 m; so far known only from Osa Peninsula. Costa Rica to Peru and Brazil.

TL 74 mm; HW 46 mm. **Identification:** Very similar to Amazon Red Darner (p. 185), but no hint of reddish in wing veins, no T-spot or brown dashes on front of thorax in either sex. Mature males entirely orange to bright red-orange. Thorax green in younger males, with age becoming even redder than mature male Amazon Red. At close range or in hand, note prominent genital lobes of male; not prominent in Amazon Red. Females with reddish brown eyes and green thorax, but older individuals can become dull brownish. Along with Amazon Red and Icarus, both sexes lack prominent T-spot present on green pilot darners. **Habitat and Behavior:** Males fly back and forth in beats that average 6 m in length, 2–3 m over ponds and grassy marshes surrounded by forest. Females oviposit in emergent and floating vegetation. Mating often takes place away from water; after mating pairs separate, females roam across breeding habitat ovipositing at many sites. Foraging individuals are at times common and, especially females, range far and wide over open areas. **Range:** Both slopes, to 700 m; not yet documented from Pacific slope south of Guanacaste. Southern Mexico to Ecuador and Venezuela.

male

female

Icarus Darner *Coryphaeschna apeora*

TL 83 mm; HW 53 mm. **Identification:** Distinctly larger than other pilot darners in CR, recognizable by prominent pair of conspicuous brown spots or short stripes on front of thorax as well as contrast between green thorax and reddish-brown abdomen in mature individuals. Thoracic markings more conspicuous than in any other pilot darners. Male eyes and face green, female eyes green with blue above. Immature coloration undocumented. **Habitat and Behavior:** Males seen at forested swamps and slow streams, which may be breeding habitat. Forages high in air, rarely comes low enough to be captured or photographed. Seemingly rare everywhere it occurs. **Range:** Northern Pacific and Caribbean slopes, to 600 m (not yet found southeast of Guanacaste and Heredia provinces). Texas to Costa Rica, Cuba.

male

female

TL 78 mm; HW 46 mm. **Identification:** No other large CR darner has a mostly green thorax striped broadly with brown, contrasting with mostly black abdomen with fine green streaks. Considerably larger than Highland and Turquoise-tipped Darners (pp. 190-191), which have brown thorax striped with green; also larger than three-spined darners (p. 174), which have green thorax with narrow brown stripes. Male eyes green to blue-green, blue in mature females. Female cerci short, quite different from pilot darners. **Habitat and Behavior:** Breeds in marshy ponds in open or at forest edge, less commonly in pools of small spring-fed streams. Males fly and hover a meter or less over open water and vegetation, often with limited patrolling area no wider than 10 m in extent. Mating often takes place up in trees. Females arrive at water and move around slowly, with much hovering low in marsh vegetation, perhaps looking for potential predators; finally land to oviposit. Both sexes hang up in shrubs and trees near water, not within forest. **Range:** Both slopes, 600–1500 m. Arizona to Argentina.

TL 71 mm; HW 47 mm. **Identification:** Only CR darner with broad chartreuse thoracic stripes in both sexes. Abdomen relatively short and thick, finely ringed with either green or blue. Heavier-bodied than neotropical darners (*Rhionaeschna*), which may also fly up and down streams, especially Highland (p. 190) and Turquoise-tipped (p. 191), both of which have narrower and slightly crooked greenish thoracic stripes in contrast with the broad, straight stripes in Mountain Stream, looking something like dabs of paint. The other two also have prominent T-spots on frons, Mountain Stream with entirely brown face. Note: assignment of the species to this genus is questionable, still under study. **Habitat and Behavior:** Rarely seen species that breeds in small streams in primary montane forest. Females observed ovipositing in stems of herbs and shrubs 1 m or more above water and up to 1–2 m back from stream margin. **Range:** Pacific slope, 1400–1650 m. Highlands of central Mexico to Panama.

male

female

NEOTROPICAL DARNERS   genus *Rhionaeschna*

Three species. This large neotropical genus is the tropical equivalent of the similarly diverse genus *Aeshna*, the most familiar darners in much of the northern hemisphere. A prominent T-spot, green- or blue-striped thorax, and brown abdomen patterned with blue or green characterizes all Costa Rican species.

Highland Darner   *Rhionaeschna cornigera*

TL 63 mm; HW 41 mm. **Identification:** Small darner with green-striped thorax and sparsely blue-marked abdomen, including blue underside of S9. Very similar to Turquoise-tipped Darner, but reaching somewhat higher elevation. Frontal thoracic stripes distinctly broader in Highland and straight; narrower and curved inward in middle in Turquoise-tipped. Lateral thoracic stripes also average wider in Highland. Green marking on side of S1 in Highland expands below, ending up over twice as wide as very narrow similar marking on Turquoise-tipped. In female Turquoise-tipped, intact cerci distinctly longer, as long as S8–10; in Highland, cerci would only reach middle of S8 if turned forward. See also somewhat similar Chartreuse (p. 170) and Mountain Stream (p. 189) Darners. **Habitat and Behavior:** Marshy ponds with much vegetation or small slow-flowing streams in or out of cloud forest. Males fly low along water edge searching for females. **Range:** Pacific slope, 1200–2300 m; on Caribbean slope, locally at 800–1300 m. Highlands of southern Mexico to Argentina.

male

female

TL 60 mm; HW 40 mm. **Identification:** Small brightly colored darner of uplands. Blue patch under S10 found only on this species and similar Highland Darner, and careful comparison necessary to distinguish them. Not much like any other CR species but compare with somewhat similar Chartreuse Darner (p. 170). **Habitat and Behavior:** Open, often marshy ponds or slow streams, usually associated with forest, where individuals roost when not at water. **Range:** Pacific slope and Central Valley, 1000–1500 m; surprisingly, only one record from Caribbean slope, 650 m. Southwestern US to Argentina, West Indies.

male

female

TL 68 mm; HW 43 mm. **Identification:** Only darner with body, face, and eye color bright sky-blue in males; all others mix of blue and green. Most of abdomen looks quite dark in contrast with blue markings on thorax and abdomen base. Male cerci forked in side view, unique in region. Females polymorphic; eyes and pale markings either blue or yellow-green; also distinguished from all other regional darners by narrower pale thoracic stripes. **Habitat and Behavior:** Large to small ponds, from entirely open to densely vegetated with grasses and sedges. Males fly back-and-forth beats over vegetation or closely along edge of open ponds. Females oviposit in floating and emergent vegetation, submerging abdomen but not wings. **Range:** Both slopes, 1100–3300 m; foraging individuals up to highest point on road over Talamanca Mountain Range. Highlands of northern Mexico to Panama.

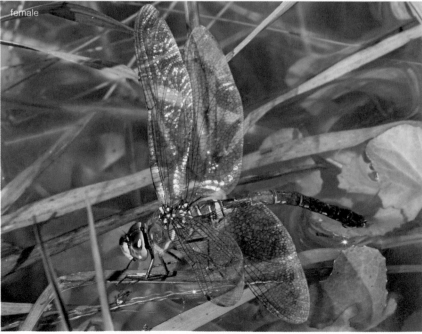

GREEN DARNERS genus *Anax*

Two species. This genus contains large darners with a green thorax and a pale-spotted abdomen with swollen base, different from all others in the region. They appear to be stronger fliers than other tropical darners and somewhat more robust in thoracic bulk. All are species of open ponds, usually perching in trees when away from water. The female cerci are short and do not break off as they get older, as happens in many other darners.

Amazon Darner *Anax amazili*

TL 70 mm; HW 47 mm. **Identification:** Sexes quite similar; only darner in region with green thorax, white- or blue-spotted abdomen, and prominent black spot surrounded by blue on top of frons. Whitish to blue ring visible from side at extreme abdomen base also distinctive. Blue-spotted Comet (p. 194) shows blue abdominal spots, lacks black marking on frons. **Habitat and Behavior:** Females at open ponds and marshes, ovipositing in low emergent or floating plants of all kinds. Such females often in dry marshes prior to rains. Males not seen cruising over water, so presumably mating away from it. Both sexes forage far and wide, recognizable by swift, wide-ranging flight when seen in evening darner flights. Rests hanging from branches in trees, sometimes high. Known to be migrant in some areas, numbers appearing at places such as mountain passes, where no breeding occurs; this not often observed in CR. **Range:** Both slopes and Central Valley, to 1500 m. Texas, Florida, and West Indies to Argentina.

male

female

Blue-spotted Comet Darner   *Anax concolor*

TL 72 mm; HW 45 mm. **Identification:** No other darner in region with green thorax and blue-spotted abdomen in both sexes. Face bright green. Amazon Darner (p. 193) has pale spots on abdomen, but in that species spots large and whitish, one per segment; S1 pale blue to whitish; and black triangle prominent on top of frons. **Habitat and Behavior:** Males fly lengthy beats over open water of ponds; females rarely seen, oviposit in emergent vegetation. Known to hang up in trees. **Range:** Both slopes, to 1500 m. Texas and West Indies to Argentina.

male

female

Clubtails (family Gomphidae)

Thirty-eight species. Gomphids can be recognized by their relatively small and well-separated eyes, usually blue or green. Patterns are cryptic, with striped thorax and banded or spotted abdomen. Males of most of the larger species have some indication of a "club," a widened subterminal part of the abdomen; the club is less developed in females of most species and not evident at all in some. Although quite diverse, as a group they are not very conspicuous in the tropics. Most breed in streams and rivers, where they may be uncommon even in optimal habitat. Their apparent rarity may have to do with the fact that some species do not occur at water except when females come to oviposit. While at times found in sunny clearings away from water, they may spend most of their time in the forest canopy. For these species, the larvae can often be found easily even where adults are not seen at all. Others are territorial at the water, much like skimmers, and are more likely to be seen.

Those most often observed on open streams are the large forceptails (*Aphylla*), leaftails (*Phyllogomphoides*), and the smaller sanddragons (*Progomphus*), the latter typically at sandy streams. Male sanddragons usually have flattened, white-tipped cerci visible at a distance. Species of knobtails (*Epigomphus*) are characteristic of smaller forest streams. They show no club, but instead the end of the abdomen is enlarged in males to enclose the rather substantial musculature of the terminal appendages. All look about the same but differ markedly by the structure of those appendages. Like many other clubtails, they show a conspicuous yellow ring on segment 7. Some clubtails are tiny (*Archaeogomphus*, for example), no more than 35 mm in length.

A prominently striped thorax is a common feature of clubtails, and those stripes are often used as identification marks. Thus, naming them is essential. We recognize five stripes on each side of the thorax: *first frontal* on either side of the midline of the front; *second frontal* at the junction between the front and side; and *first, second, and third lateral* along each side. These stripes can be light on a dark background or dark on a light background, and it is their presence or absence that provides identification marks. In cases in which there might be confusion, more information is given.

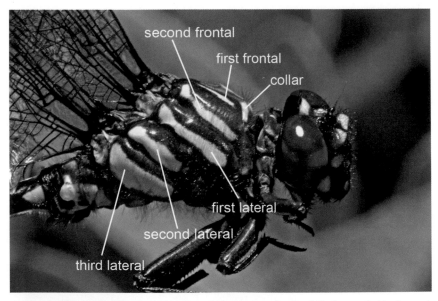

Names used for the thoracic stripes in Gomphidae (a reared male of *Phyllogomphoides burgosi*, La Selva Biological Station).

LARGE FORCEPTAILS genus *Aphylla*

Three species. These large clubtails are found on streams, as well being the only ones commonly occurring at sloughs and ponds. Males have a relatively narrow club, flanges on S8, forceps-like cerci, and barely visible epiproct. They are the same size as leaftails and sometimes occur with them; males have very different appendages, but females could be confused with leaftails. Note that the female cerci are mostly white, contrasting with a black abdomen tip in leaftails, not so in forceptails. Though awkward, "forcepstail" would have been a more linguistically correct common name.

Broad-striped Forceptail *Aphylla angustifolia*

TL 62 mm; HW 36 mm. **Identification:** Males distinguished from other forceptails by black-and-white look: thorax vividly striped black and cream; abdomen black with cream spots from base to S7, where especially prominent and distinctive, then becoming increasingly reddish on sides of dark segments at rear. Club rather narrow in male, absent in female. Male cerci whitish inside, distinctive for this species. Most similar to Ringed Forceptail (p. 199), which is substantially smaller and has prominent club, S8 wide and reddish with black-edged convex flange, and dark cerci. S8 overall dark and flange straight-edged in Broad-striped. Females look quite alike, but first frontal stripe on thorax broadly connected to collar stripe and pointed above in Ringed, not in contact and narrow throughout in Broad-striped. Female could be confused with leaftail except for reddish abdomen. **Habitat and Behavior:** Ponds, including those for growing tilapia, typically well vegetated but also in open farm ponds. Also on small slow streams in pastures. Males perch on twigs near water and make frequent patrol flights along pond margins and up and down streams. **Range:** Caribbean slope, to 750 m. Texas and Louisiana to Costa Rica.

male

female

TL 60 mm; HW 37 mm. **Identification:** Much more reddish and brown than Broad-striped, with broader club in male and distinct club in female, formed by wide flanges on S8. So far not found together. Thorax dark brown with narrow yellow stripes, abdomen mostly reddish with dark markings at ends of segments, darker above on S8-9. **Habitat and Behavior:** A species of slow streams; farther north, also known from ponds and lakes. **Range:** Northern Pacific slope, to 800 m. Arizona and Texas to Costa Rica.

male

female

TL 63 mm; HW 36 mm. **Identification:** Overlaps with other two large forceptails. Thorax broadly striped dark brown and pale greenish like others, distinguished by mostly dark brown abdomen with narrow reddish to yellowish flanges on S8-9. **Habitat and Behavior:** Males guard small pools in slow streams, alternating with periodic patrolling. Also seen at entirely open ponds. **Range:** Both slopes, to 200 m. Southern Mexico to Venezuela.

male

female

SMALL FORCEPTAILS   genus *Phyllocycla*

Two species. This genus is closely related to the large forceptails, but a variety of structural characters distinguish the two genera. *Phyllocycla* has species that look very much like species of *Aphylla* but also many smaller species with different appendages and different appearance. They are committed to stream habitats, not living in ponds as some large forceptails do. The two genera have in common the very long 10th segment of the larval abdomen, which is used as a breathing tube to reach the water for respiration when the larva is buried in mud.

Ringed Forceptail   *Phyllocycla breviphylla*

TL 56 mm; HW 32 mm. **Identification:** Fairly large clubtail with brown, yellow-striped thorax and black abdomen prominently ringed with yellow at bases of S3–7, slightly more yellow on S7. Distinctly expanded S8-9, reddish with black margins. Male cerci curved inward as in *Aphylla* but tip bent downward and sharply pointed. No epiproct visible. Rather similar to Broad-striped Forceptail (p. 196) but smaller, with larger club, and less yellow on S7. **Habitat and Behavior:** Typically on slow streams with wooded banks. When perched, abdomen hangs downward. **Range:** Far northern Caribbean slope, to 300 m; known from near Santa Cecilia east to Los Chiles. Texas to Costa Rica.

male

female

199

TL 47 mm; HW 26 mm. **Identification:** Small clubtail with brown, yellow-striped thorax and yellow rings at base of S3–7. First frontal stripe pointed above and wide below, no apparent collar stripe. Male with expanded S7-8, reddish with black edges. Short, thick incurved male cerci flattened and toothed at ends, no apparent epiproct. **Habitat and Behavior:** Slow-flowing streams and rivers, especially with sandy bottom. Both sexes perch on leaves, often above head height, resting almost vertically on hanging leaves. **Range:** Pacific and Caribbean slopes, to 100 m. Eastern Mexico to Panama.

male

female specimen

LEAFTAILS   genus *Phyllogomphoides*

Five species. Leaftails are about the size of forceptails but much darker, with yellow, cream, or pale blue stripes on a dark brown thorax; a black abdomen with white or yellow spots on the sides of the segments; and a prominent white to cream band at the base of S7. The abdomen is clubbed only in males, and the common name comes from species such as Horned Leaftail with wide leaflike projections on S8, S9, or both. The male cerci are long, curved inward, and expanded at the tip into blunt ends. The epiproct is short but visible from the side and deeply forked. Females hover over forest streams while dropping eggs singly into the water, often in spots with eddies or deep riffles, not a common behavior in dragonflies but known also in other clubtail genera.

Panamanian Leaftail   *Phyllogomphoides appendiculatus*

TL 54 mm; HW 32 mm. **Identification:** One of two species of leaftails that have only two pale lateral stripes on thorax, the middle one absent. In both species the second frontal stripe is reduced to a spot or may be absent. See Two-striped (p. 202) for distinction from that species. **Habitat and Behavior:** Slow-flowing streams in forest. **Range:** Pacific slope, to 200 m; from Puntarenas area south. Costa Rica and Panama.

male

female

TL 54 mm; HW 32 mm. **Identification:** One of two leaftails with only two pale stripes visible on each side of thorax; compare with very similar Panamanian (p. 201). Males probably impossible to distinguish except in hand or with very good photos. In dorsal view, male cerci smoothly rounded as they curve inward at dorsal tooth (slightly angled inward in Panamanian). Epiproct notched to about half its length, directed straight backward (slightly shorter but more deeply notched in dorsal view, directed somewhat upward in Panamanian). In side view, Panamanian male cerci slender in middle with mid-dorsal tooth sharply defined, cerci of Two-striped wider with dorsal tooth barely visible. Lobes of female subgenital plate rounded in Two-striped, angled in Panamanian. **Habitat and Behavior:** Slow-flowing areas of rivers and streams down to quite small ones, often in pastures. Males perch in sun over and near water. Both sexes frequently found along a road through hill forest well away from water. **Range:** Pacific slope, to 400 m. Common in Guanacaste and appears to overlap with Panamanian in Puntarenas area but needs more study. Southern Mexico to Costa Rica.

male

female

TL 58 mm; HW 38 mm. **Identification:** As in Horned (p. 205), a complete complement of stripes in both sexes. Stripes somewhat wider in larger Horned, and male Horned also has much larger "leaf" on S8. **Habitat and Behavior:** Small forest streams. **Range:** Caribbean slope, to 100 m. Eastern Mexico to Costa Rica.

male

female

TL 62 mm; HW 38 mm. **Identification:** Much like Horned and Tuxtla (p. 203) Leaftails, with three pale stripes on each side of thorax in both sexes, but second frontal stripe on each side represented by only a dot at upper end. **Habitat and Behavior:** Slow-flowing areas of rock-bottomed rivers and streams in forest. Females seen ovipositing in riffles and eddies behind large rocks. **Range:** Northern Pacific slope, 300–1100 m. Northern Mexico to Costa Rica.

immature male

female

TL 67 mm; HW 41 mm. **Identification:** Largest leaftail in CR, three lateral stripes all fairly wide and males with lobe-like posterior projections on each side of S8. Cerci with long ventral spur near base; epiproct sharply curved upward with short narrow slit between branches. Females distinguished by same arrangement of thoracic stripes and slightly larger size than other leaftails, plus pair of acute horns on top of frons. **Habitat and Behavior:** Slow-flowing streams in forest. **Range:** Caribbean slope, to 600 m. **Costa Rican endemic.**

male

**genus *Phyllogomphoides* (p. 201)
male appendages in lateral and dorsal views**

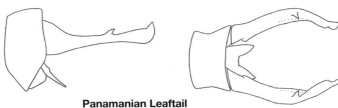

Panamanian Leaftail
Phyllogomphoides appendiculatus (p. 201)

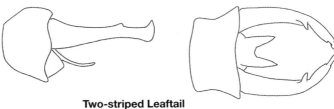

Two-striped Leaftail
Phyllogomphoides bifasciatus (p. 202)

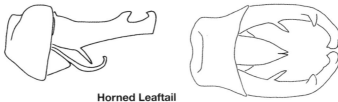

Horned Leaftail
Phyllogomphoides burgosi (p. 205)

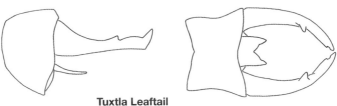

Tuxtla Leaftail
Phyllogomphoides pugnifer (p. 203)

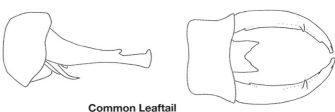

Common Leaftail
Phyllogomphoides suasus (p. 204)

SANDDRAGONS genus *Progomphus*

Five species. These are brightly marked clubtails that include the smallest members of the family and two larger species that match the size of ringtails and the even larger knobtails. As in most other gomphids, the thorax is vividly striped and the abdomen mostly black with a conspicuous pale ring on the base of S7. The male cerci are fairly long and flattened, with white tips. The epiproct is long, deeply incised as if two-parted, and with additional teeth or lobes near the tip of each branch. The breeding habitats are sandy or silty areas of streams and rivers, as the larvae are highly adapted for burrowing in sand. The two larger species are often seen at their breeding habitat, beachlike stream margins and sand bars, but not so with the smaller ones, which probably spend little time there except to lay eggs. Females have been observed laying eggs along the sandy edges of streams and rivers.

Zebra-striped Sanddragon *Progomphus clendoni*

TL 50 mm; HW 29 mm. **Identification:** One of two largest sanddragons, with dark and light stripes on thorax about equal width; mostly black abdomen patterned with cream-white, including conspicuous ring on S7. Basal segments may each have a long, pointed pale dorsal streak, broad at base and flared again at tip, in some individuals extending to middle segments. See Mexican Sanddragon (p. 208). **Habitat and Behavior:** Streams and small rivers. Males perch on gravel or sand beaches, emergent rocks and vegetation, more often flat on substrate than smaller sanddragons. **Range:** Pacific slope, to 850 m; Caribbean slope, to 600 m. Northern Mexico to Panama.

male

female

TL 51 mm; HW 26 mm. **Identification:** Size and general appearance of Zebra-striped (p. 207) but more reddish on sides of S8–10 in both sexes. Pale markings on middle abdominal segments (S3–6) average broader, presenting more ringed look. Appendages distinguish males readily. Cerci black with white terminal third in Mexican, slightly shorter and mostly white in Zebra-striped. In Mexican split epiproct curves sharply inward at end of each branch, while in Zebra-striped each branch remains straight; in both species an outer tooth directed upward. Female Mexican with pair of depressions on either side of vertex adjacent to eyes, female Zebra-striped with low tubercles in same location. **Habitat and Behavior:** Small sandy or muddy rivers, slower flowing than those where Zebra-striped occurs. Behavior seems identical to Zebra-striped, perching both on sand and low plants. Several males alternated patrolling a stream and perching next to water. **Range:** Pacific slope, to 100 m; known only from Guanacaste. Western Mexico to Costa Rica.

male specimen

TL 36 mm; HW 22 mm. **Identification:** Largest of three small sanddragons, the others Forest (p. 210) and Pygmy (p. 211). Differs from others in having only two pale lateral stripes on thorax, middle stripe missing. Also lacks spot representing reduced second antehumeral stripe present on front of thorax in Forest. Stigmas shorter than in Forest. **Habitat and Behavior:** Small to medium forested streams. Larvae surprisingly common and widespread considering adults so rarely seen; most likely to be found as a recently emerged teneral. **Range:** Pacific slope, 200–800 m; Caribbean slope, 500–1300 m. Costa Rica to Peru.

male

female specimen

TL 34 mm; HW 19 mm. **Identification:** One of three small sanddragons, distinguished from Anomalous (p. 209) and Pygmy by thoracic pattern, three wide pale lateral stripes and second frontal stripe represented by tiny spot at upper end. Stigmas elongate in Forest, more than half the distance from stigma to nodus (about half in other two species). **Habitat and Behavior:** Streams and small rivers. Quite common on Pacific slope in Monteverde region. Two females oviposited at margin of sandy stream with moderate current. **Range:** Pacific slope, 200–850 m; also on Caribbean slope, at 500 m near Braulio Carrillo NP. Southern Mexico to Costa Rica.

immature male

female

TL 33 mm; HW 18.5 mm. **Identification:** Tiny sanddragon, about size of Forest and only slightly smaller than Anomalous (p. 209). Differs from them in lacking spot representing second frontal stripe that is present (presumably always) in Forest, and has three pale lateral stripes (two in Anomalous). As small as Mexican Snout-tail (p. 232), differs from that species in mostly black abdomen (mottled and barred in Snout-tail), much longer stigmas, and quite different appendages. **Habitat and Behavior:** Small rocky streams and rivers. Males cruise within few centimeters of water surface, presumably searching for females, and blend well with rippled water. Female seen to drop eggs in stream while hovering less than a meter above it. This oviposition style may be more common among tropical clubtails than is currently thought. Several females observed in Ecuador oviposited in a sandy river edge with swarm of males buzzing around them. **Range:** Pacific slope, to 600 m; Caribbean slope, at 850 m. Less commonly seen than other two small sanddragons. Eastern Mexico to Bolivia and Brazil.

male

female specimen

RINGTAILS genus *Erpetogomphus*

Six species. Ringtails have their greatest abundance and diversity farther north, and all Costa Rican species are local and uncommon, but they can be easily recognized as the only clubtails with thorax striped with bright green (even jade or turquoise). The extent of green on the thorax, from narrow stripes to almost filling it (in which case it looks green with brown stripes), is important for species identification. In males, S8–10 are enlarged into a small club, with S10 especially large to hold the large appendages; there is no such enlargement in females. Male cerci distinguish each species, male epiproct short and strongly upcurved in all species but varies in length. Thoracic color pattern of females is generally like that of their respective males but less distinct in some. A close look at the area around the vertex and ocelli, as well as the shape of the subgenital plate, may be necessary to distinguish females of similar species. The first three species have pale or mostly pale faces, the remainder dark faces.

One-striped Ringtail *Erpetogomphus bothrops*

TL 48 mm; HW 27 mm. **Identification:** One of two ringtails with pale face and thorax largely green, with only faintly marked brown lateral stripes. Differs from Lime Ringtail in downcurved male cerci (straight in Lime). Both sexes of One-striped have obvious dark median thoracic stripe, poorly indicated or absent in the very lightly marked Lime. **Habitat and Behavior:** Forested and semi-open rocky streams. Males perch on rocks and twigs at water. **Range:** Northern Pacific slope, to 100 m; known only from Guanacaste. Northern Mexico to Costa Rica.

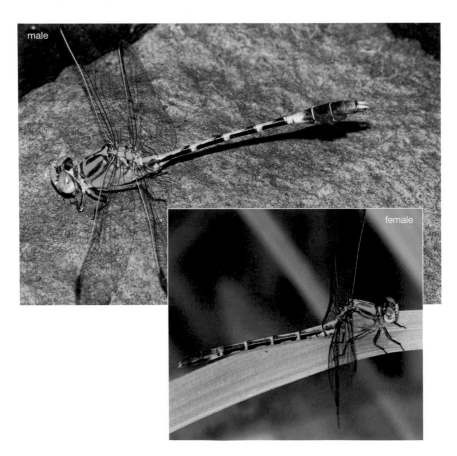

TL 44 mm; HW 26 mm. **Identification:** Both sexes most lightly marked of CR ringtails, with pale face and thorax, dark thoracic stripes somewhat obscure, and no stripes on sides. Also only species in which males have long, straight, pointed cerci with no teeth. Females have much-divided sub-genital plate with widely separated, long parallel branches. **Habitat and Behavior:** So far found only at open marshes, in one instance with a ditch flowing through it. Presumably breeds in running water. **Range:** Known only from the Central Valley, 1400–1700 m. Guatemala to Costa Rica.

male

TL 41 mm; HW 24 mm. **Identification:** Face pale with dark markings, thoracic stripes somewhat narrower than three dark-faced species (but broader than in photos, which are from Texas); male appendages most like those of very different One-striped. Face blue in more northerly populations, color not described in CR. Females lack ocellar groove of Knob-tipped and Tristan's (p. 217), also lack ridge at rear of vertex; compare with slightly larger Schaus's (p. 216). **Habitat and Behavior:** Forested and open small to medium-sized streams. Males perch on vegetation along shore and in nearby clearings up to several meters above ground. **Range:** Northern Pacific slope and La Selva Biological Station on Caribbean slope, to 600 m. Texas to Costa Rica.

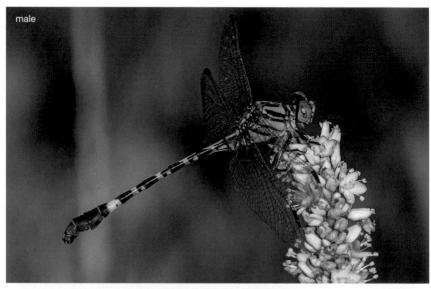

male

female

TL 41 mm; HW 27 mm. **Identification:** One of three dark-faced ringtails with heavily striped thorax, brown stripes wider than green. Males distinguished from all other ringtails by downcurved cerci with prominent dorsal tooth at two-thirds length. Only Knob-tipped and Schaus's (p. 216) have green thoracic stripes so narrow that thorax looks more brown than green. Females not quite so heavily striped, difficult to distinguish in all species of this group. However, females of two of them (Knob-tipped and Tristan's [p. 217]) have very deep groove between lateral ocelli, completely hiding median ocellus from above. Groove narrower in Tristan's, with convex sides, while sides straight in Knob-tipped. The groove corresponds to short, thick, narrow-tipped epiproct of males of these species, fitting together like lock and key during mating. Similarly colored Blue-faced and Schaus's have no such groove. **Habitat and Behavior:** Small to medium-sized rocky rivers with fast flowing riffles. Males perch on dead twigs and leaves at edge of, or over, water. **Range:** Lowlands of northern Pacific Slope, around to La Fortuna on Caribbean slope, to 700 m. Eastern Mexico to Costa Rica.

male

female

TL 45 mm; HW 29 mm. **Identification:** Darkest, most heavily marked of CR ringtails. Most similar to Knob-tipped (p. 215) and Tristan's but lacks any teeth on male cerci; other species are also at lower elevation. Cerci most similar to One-striped, a species with lightly striped thorax. Both Schaus's and Knob-tipped have green second frontal thoracic stripe represented by spot at upper end separate from very narrow fine line below, while Tristan's has first frontal stripe curved outward to include that spot. As in somewhat similarly colored female Blue-faced, lacks ocellar groove, but has pronounced ridge bordering rear of vertex, no ridge in Blue-faced. **Habitat and Behavior:** Small rocky and sandy streams in cloud forest. Males come down from canopy briefly to investigate quieter pools, fly along the shoreline for few seconds, occasionally perching on stones, then return straight up to treetops. **Range:** Pacific slope, 1400–1600 m; Caribbean slope, 800–1600 m. Guatemala to Costa Rica.

male teneral

female

TL 43 mm; HW 27 mm. **Identification:** Dark face and well-defined dark thoracic stripes. Males differ from similar species in having straight, short cerci with both dorsal and ventral teeth very prominent in side view, making it rather thorny looking, and also having very short epiproct. **Habitat and Behavior:** Streams and rivers. Males perch in sun at edge or over water, and females are in nearby clearings. **Range:** Pacific slope, to 600 m. Costa Rica and Panama.

male specimen

teneral female

genus *Erpetogomphus* (p. 212)
male appendages in lateral and dorsal views

One-striped Ringtail
Erpetogomphus bothrops (p. 212)

Knob-tipped Ringtail
Erpetogomphus constrictor (p. 215)

Lime Ringtail
Erpetogomphus elaphe (p. 213)

Blue-faced Ringtail
Erpetogomphus eutainia (p. 214)

Schaus's Ringtail
Erpetogomphus schausi (p. 216)

Tristan's Ringtail
Erpetogomphus tristani (p. 217)

KNOBTAILS   genus *Epigomphus*

Thirteen species. Knobtails are medium-sized, slender clubtails of forest streams occurring throughout the country. They lack a club but have the terminal abdominal segment enlarged in males to hold the rather massive appendages. All species can be distinguished by the shape of those appendages—both cerci and epiproct are important—but otherwise identification characters are few and far between, and even less evident in females. Close-up photos of female heads from above can be very helpful. Note that females of three species are unknown, so that distinctions made among the species do not take them into account. Some identifications can be made on the basis of location, as only a small subset of species occurs in most places. Armed, Cloudforest, Camel, Plate-crowned, Fork-tipped, Common, and Lowland are all common at one site or another, but the other six species remain poorly known.

Some species have a complete second frontal thoracic stripe, others lack the stripe, with a spot where the upper end of the stripe would be; in Cloudforest and some Plate-crowned, the lower end of the stripe is also represented as a streak. Spotted species are listed first, followed by the two species with the incomplete stripe, and then the fully striped species. At several sites where two species occur together, one is striped and one is spotted, thus facilitating field identification.

Males generally hold territories while perched on leaves or twigs at the edge of streams and pursue any male or female that approaches their site. A few species oversee their territory from perches in trees above the water. Most females are seen when they come to water to oviposit, which they do alone while flying slowly over the surface and dipping the abdomen tip or—unique among Costa Rican dragonflies—perching on the ground and inserting their eggs into wet mud at the water's edge.

Camel Knobtail   *Epigomphus camelus*

TL 55 mm; HW 36 mm. **Identification:** Spotted. Male epiproct extends slightly beyond rather short cerci, each branch ending in sharp point with upward-directed spine coming from it. No other species shows that in lateral view. Note "hump" on S10 that gives it its name. Female has pair of tubercles above occipital ridge, large and no farther apart than width of each tubercle. Female Common (p. 226), with which it occurs, also has tubercles but smaller and farther apart, and lacks thoracic spots. **Habitat and Behavior:** Small forest streams, waterfalls, and swampy trickles. In one observation, males perched on leaves and rotten logs, hovered from spot to spot before landing. **Range:** Caribbean slope, 850–1200 m. **Costa Rican endemic.**

male

TL 54 mm; HW 35 mm. **Identification:** Spotted. Male cerci gently curved downward at tip, with series of jagged bumps on undersides. Epiproct deeply notched, with branches curved inward and pointed at tip. Female unknown. Should overlap with Armed, Horned, and Lowland, and of those only Lowland has thoracic spots. **Habitat and Behavior:** Nothing known, presumably on forest streams. **Range:** Caribbean slope, to 100 m. Type locality near Siquirres. **Costa Rican endemic.** Not illustrated.

Lowland Knobtail *Epigomphus tumefactus*

TL 50 mm; HW 34 mm. **Identification:** Spotted. Male cerci in dorsal view arched outward, then bent inward; broad in lateral view with row of tiny teeth under tips. Epiproct shorter than cerci, wide and shallowly notched, each branch upturned at tip with a small tooth. Females with shallow grooves on either side of ocelli and pair of tubercles adjacent to eyes behind occipital ridge. Occurs with Armed (p. 225) in lowlands and Common (p. 226) at higher elevations, distinguished from Armed by shape of appendages and from Common by spotted thorax. **Habitat and Behavior:** Spring seeps and small silt- or rock-bottomed streams in forest, where territorial males guard pools of slow moving water. Immatures of both sexes perch in sunny light gaps along forest trails well away from water. Most common and widespread knobtail in lowlands of both slopes except south Pacific and Osa Peninsula, where Fork-tipped is the common species. **Range:** Northern Pacific and Caribbean slopes, to 1200 m. **Costa Rican endemic.**

male

female

Fork-tipped Knobtail *Epigomphus quadracies*

TL 49 mm; HW 32 mm. **Identification:** Spotted. Male cerci broad and truncate, inner corner longer, curved notch between cerci and corner producing pronounced forked appearance in lateral view. Epiproct with deep rounded notch, pair of upward pointing teeth at end of each branch. **Habitat and Behavior:** Small streams in forest. **Range:** Pacific slope, north to Santa Rosa NP, to 100 m. Guatemala to Panama.

male

female

TL 49 mm; HW 32 mm. **Identification:** Spotted. Male cerci truncate from above, tip curled under sharply and hooklike from side. Epiproct longer than cerci, notched to about half length and with two short upright apical teeth on each branch. **Habitat and Behavior:** Spring seeps and small pools in forest streams, which males guard from perches on leaves. **Range:** Pacific slope, from Central Valley to Guacimal near Monteverde, 300–900 m. **Costa Rican endemic.**

male

TL 49 mm; HW 35 mm. **Identification:** Spotted. Male cerci slender in dorsal view, convex on outer surface and concave on inner surface, thus slightly curved; short ventral tooth at half-length not always visible in lateral view. Epiproct deeply divided with U-shaped notch, slender branches parallel and slightly notched at tips. Long, curved spur directed forward and up at base of each branch, diagnostic but usually hidden by cerci. Females have pair of horns or tubercles posterior to ocelli, rather like Cloudforest (p. 224) and Horned (p. 226) but shorter and with rounded tips. **Habitat and Behavior:** Forest streams. **Range:** Caribbean slope, 700–800 m. Type locality Tuis, near Turrialba. **Costa Rican endemic.** Not illustrated.

TL 56 mm; HW 36 mm. **Identification:** Outer frontal thoracic stripe broken in middle, shared only with Cloudforest (p. 224). Note some Plate-crowned with lower part of stripe very poorly developed or lost with age, could look spotted. Male cerci truncate from above, inner corner a long point; in lateral view, that point visible, resembling an inner tooth. Notch in epiproct broad and rounded, pair of upward-pointing short teeth at end of each branch. Flat plates extending rearward from ocelli in female unique. **Habitat and Behavior:** Forest streams with silt, gravel, or rocky bottom. **Range:** Northwest Pacific and Caribbean slopes, 500–1200 m; Guanacaste and Tilarán Mountain Ranges south to Braulio Carrillo NP. **Costa Rican endemic.**

teneral female

TL 53 mm; HW 38 mm. **Identification:** Outer frontal thoracic stripe broken in middle, as in Plate-crowned (p. 223). Cerci short and thick, incurved and downcurved at tips with two barely perceptible subapical teeth. Epiproct longer than cerci, deeply divided into two upward curved branches reminiscent of manta ray "horns." Highly distinctive male appendages and thoracic-stripe pattern distinguish it from Common (p. 226), with which it occurs. Females have pair of long pointed horns behind ocelli, duplicated only in Horned, with which no overlap. **Habitat and Behavior:** Small streams in cloud forest. Larvae common, adults rarely seen. Males perch on rocks and leaves near water. One female oviposited directly into wet mud at water's edge, others at edges of small rocky riffles in stream. **Range:** Tilarán and Central Mountain Ranges, 1400–1700 m. **Costa Rican endemic**.

male

female

TL 49 mm; HW 33 mm. **Identification:** Striped. Of Caribbean lowland species, this has male cerci most strongly bent downward in side view, with tips pointing down like overhanging fangs. Epiproct much shorter, hidden between cerci, and shallowly notched with two short teeth at tip. Females have short, rounded tubercles behind lateral ocelli. **Habitat and Behavior:** Small forest streams and spring seeps. Males guard areas of clear, shallow water over silt bottom, including springheads. Often found in same areas as Lowland Knobtail (p. 220), a spotted species. Markings on abdomen pale yellow in Lowland, slightly darker and orange-tinged in Armed. **Range:** Caribbean slope, to 700 m. **Costa Rican endemic.**

male

female

TL 54 mm; HW 34 mm. **Identification:** Striped. Male cerci longer than epiproct, wide and curved downward. Epiproct very wide, not notched at all but with pair of median tubercles that are unique. Females have a pair of short divergent horns behind ocelli, shorter than in Cloudforest (p. 224) and with pointed tip; no overlap in range. **Habitat and Behavior:** Small forest streams. **Range:** Southern Caribbean slope, to 100 m. Type locality Suretka. **Costa Rican endemic.** Not illustrated.

Common Knobtail *Epigomphus subobtusus*

TL 49 mm; HW 32 mm. **Identification:** Striped. Male cerci gradually downcurved in distal half, relatively narrow. Epiproct with very deep notch forming acute angle, male recognized by openings between appendages conspicuous in either dorsal or lateral view, not completely closed up as in numerous other species. Females have a pair of short upright conical peglike tubercles on occipital ridge, only duplicated in Camel (p. 219), in which tubercles larger and closer together. **Habitat and Behavior:** Forested streams and rivers, quite broad in habitat choice from tiny, swampy spring seeps to large rocky rivers. Most common and widespread knobtail at middle elevations. **Range:** Pacific and Caribbean slopes, 700–1700 m. Guatemala to Costa Rica.

male

female teneral

TL 50 mm; HW 33 mm. **Identification:** Striped. Male cerci curved downward toward ends, with series of tiny teeth under tips. Epiproct quite wide with about a 90° angle notch, each branch ending in an upward pointing sharp tooth. Female unknown. Occurs with Common, Wagner's (p. 228), Lowland, and Plate-crowned; latter two have thoracic spots, so must only be distinguished from Common and Wagner's. **Habitat and Behavior:** Only known specimen collected at shallow, partly shaded pool in spring stream at interface of forest and pasture. **Range:** Known only from type locality at Rio Guacimal, near Monteverde, at 725 m. **Costa Rican endemic**.

male

TL 47 mm; HW 31 mm. **Identification:** Striped. Cerci straight in dorsal view with truncate tips, inner corner longer; wide and curved to rounded points in lateral view. Epiproct much shorter than cerci, expanded mid-length and then pointed at tip in lateral view; in dorsal view wide and truncate, scarcely notched. Female unknown. Found with Common (p. 226), Morrison's (p. 227), Lowland, and Plate-crowned; latter two have thoracic spots, so identification based on comparing male appendages of Common and Morrison's. **Habitat and Behavior:** Only known specimen collected at shallow, partly shaded pool in spring stream at interface of forest and pasture. **Range:** Known only from type locality at Rio Guacimal, near Monteverde, at 725 m. **Costa Rican endemic.**

male

genus *Epigomphus* (p. 219)
male appendages in lateral and dorsal views

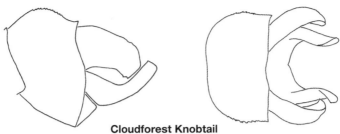

Cloudforest Knobtail
Epigomphus bosquenuboso (p. 224)

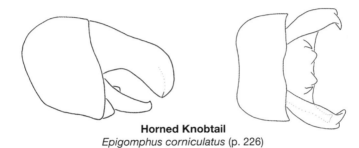

Horned Knobtail
Epigomphus corniculatus (p. 226)

Plate-crowned Knobtail
Epigomphus echeverrii (p. 223)

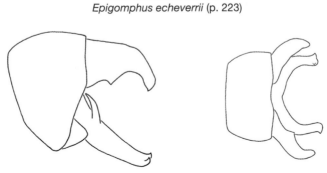

Cartago Knobtail
Epigomphus verticicornis (p. 223)

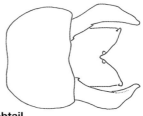

Limon Knobtail
Epigomphus houghtoni (p. 220)

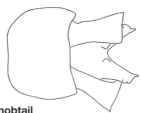

Lowland Knobtail
Epigomphus tumefactus (p. 220)

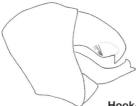
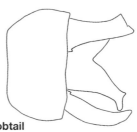

Hook-tipped Knobtail
Epigomphus subsimilis (p. 222)

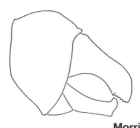

Morrison's Knobtail
Epigomphus morrisoni (p. 227)

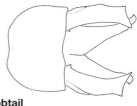

Common Knobtail
Epigomphus subobtusus (p. 226)

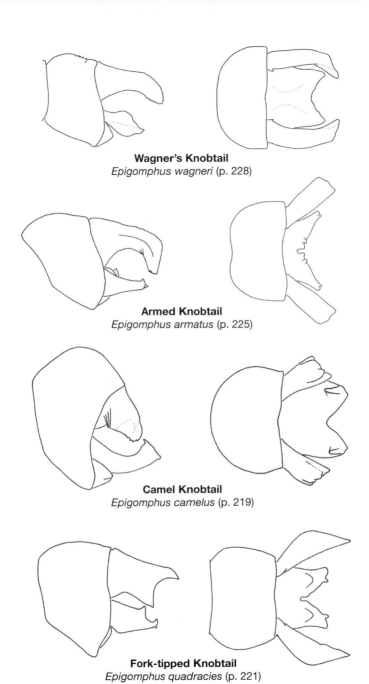

Wagner's Knobtail
Epigomphus wagneri (p. 228)

Armed Knobtail
Epigomphus armatus (p. 225)

Camel Knobtail
Epigomphus camelus (p. 219)

Fork-tipped Knobtail
Epigomphus quadracies (p. 221)

TL 35 mm; HW 19 mm. **Identification:** Among smallest of CR clubtails; abdomen slender throughout in females but enlarged into short club on males because of prominent flanges on S8. S10 with odd pair of hooks projecting up at base of segment and long flat projection extending over appendages, reminiscent of a snout. S1–6 with pale spots, S7 with larger pale marking. Separated eyes distinguish it from small skimmers such as dashers that share striped thorax and pale markings on S7. In this species, marking on S7 covers much of segment rather than pair of spots that dashers show. Females with two backward-pointing "horns" between eyes. **Habitat and Behavior:** Slow-flowing forest streams and rivers. Larvae not rare but almost never seen as adult, most likely to be encountered just after emergence. Presumably lives in canopy. **Range:** Pacific and Caribbean slopes, to 100 m. Eastern Mexico to Peru and Venezuela.

female specimen

TL 42 mm; HW 23 mm. **Identification:** Small clubtail with slender abdomen slightly enlarged toward tip, vividly patterned black with elongate white rectangles on sides of S3–6 and basal 60% of S7 white. Male cerci slightly longer than S10, sharply pointed and white, diverging outward. Compare with two other small, slender species in region. Pincertail has much smaller white markings on abdomen in both sexes, only narrow rings, and wider dark stripes on sides of thorax. Male cerci in pincertail distinctly longer than S10 and turned inward at tips. Both species with pale faces. Isthmian Pegtail has dark face, little indication of white on abdomen (tiny spots in males and narrow streaks in females) and long, straight, parallel white cerci in males. Note a somewhat similar common skimmer with blue eyes, Blue-eyed Setwing (p. 322), but eyes larger and in contact, pale markings smaller, wings longer, and abdomen shorter in Setwing. **Habitat and Behavior:** Slow flowing, mud-bottomed forest streams. Presumed territorial males perch on dead twigs close to ground at forest edge near water or even on branches over water. Nonterritorial males and females on similar perches farther into forest and small light gaps away from water. **Range:** Pacific and Caribbean slopes, to 400 m. Eastern Mexico to Panama.

male

female

TL 44 mm; HW 24 mm. **Identification:** Small clubtail with very slender abdomen, slightly enlarged toward tip. Front of thorax with pale stripe and spot on each side, while similar Slender Clubtail has two pale frontal stripes. Abdomen with narrow white rings at base of S4–7. Male cerci white, long, and incurved at tips. Epiproct short and curled upward. See Slender Clubtail (p. 233) and Isthmian Pegtail, only similar species. **Habitat and Behavior:** Very small streams to medium rivers with moderate flow. Rarely seen perched low on leaves along forest trails and in light gaps. **Range:** Pacific and Caribbean slopes, to 700 m. Nicaragua to Ecuador.

male

female

234

TL 43 mm; HW 25 mm. **Identification:** One of three small, virtually clubless clubtails in CR, all uncommon and rarely seen. Males with two pairs of pale stripes on front of thorax and tiny pale markings at bases of S3–7, pale lateral streaks on those segments in females. Male cerci white, longer than S10, straight and parallel, with small tooth at base and slightly incurved tips. See Slender Clubtail (p. 233) to distinguish the three. Might be confused with small sanddragons, but note different thoracic pattern. **Habitat and Behavior:** Streams and rivers. Male seen perched on leaves in shade of canopy. **Range:** Pacific and Caribbean slopes, to 950 m. Costa Rica to Colombia.

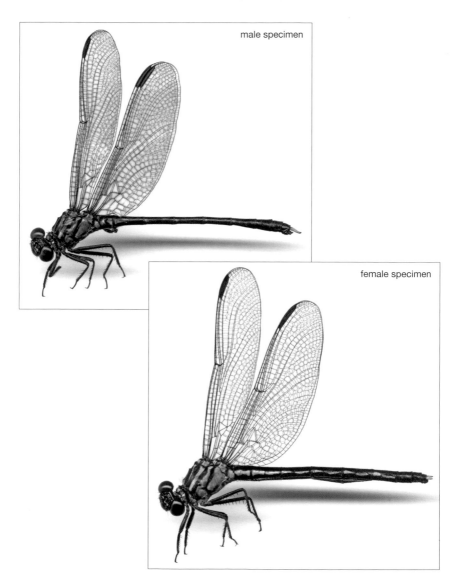

male specimen

female specimen

miscellaneous Gomphidae (pp. 232–235)
male appendages in lateral and dorsal views

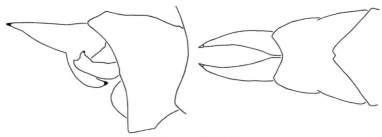

Slender Clubtail
Agriogomphus tumens (p. 233)

Mexican Snout-tail
Archaeogomphus furcatus (p. 232)

Neotropical Pincertail
Desmogomphus paucinervis (p. 234)

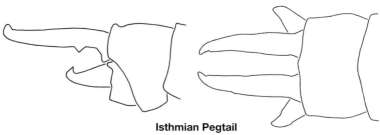

Isthmian Pegtail
Perigomphus pallidistylus (p. 235)

Spiketails (family Cordulegastridae)

One species. This is a northern group spanning the temperate parts of North America and Eurasia. The sole species in Costa Rica, which is large and black with yellow markings, lives high in mountains at 1400 meters and above. Females have a spike-like structure (a pseudo-ovipositor that is less complex than the ovipositor in darners) under the abdomen tip, with which they insert eggs into muddy substrates. Their green eyes barely touch, thus they appear more similar to clubtails than to darners and skimmers.

Apache Spiketail *Cordulegaster diadema*

TL 77 mm; HW 45 mm. **Identification:** Color pattern unique in CR, as are turquoise green eyes that just barely meet. No other species with female's "spike" at abdomen tip, but note somewhat similar structure in woodskimmers (*Uracis*, p. 271). Most similar in size and flight style to some montane darners such as Mountain Stream Darner (p. 189) but clearly distinct if eyes and ringed pattern seen. Yellow markings finer than those in same species farther north in Mexico. **Habitat and Behavior:** Breeds in small, slow streams in cloud forest, especially in pools where sandy bottom overlain by mud. Males patrol streams in slow flight searching for females. After mating, females lay eggs alone by hovering in place and repeatedly jabbing abdomen down into mud bottom just out from shore. Both sexes perch hanging down at 45° in sunny clearings and forage for insects by sallying out after them. Other spiketails take many bees and wasps, but nothing known about diet of this one. **Range:** Probably all mountain ranges, 1400–2900 m. Southwestern US to Panama.

male

immature female

Emeralds (family Corduliidae)

Three species. Abundant in temperate latitudes, emeralds are surprisingly rare and poorly known in the tropics. The three species of *Neocordulia* in Costa Rica are rarely seen, and finding one is a red letter day! They are dark metallic green with green eyes, and presumably all breed in forest streams. All are fliers, foraging in flight and then hanging up like a darner. As far as is known, they are mostly active at dusk but sometimes also forage during cloudy weather. The anal loop in the hindwings is shaped like a windsock rather than a foot as in skimmers. This genus may not belong in the temperate-zone emerald family and has been placed alternatively in the family Synthemistidae along with two other Neotropical genera; however, its classification is still in question.

Bates's Emerald   *Neocordulia batesi*

TL 50 mm; HW 33 mm. **Identification:** Emeralds easily distinguished from all other regional dragonflies by bright green eyes (brown in immature), metallic green thorax, and black abdomen. See other emerald species for differences within group. Male abdomen weakly expanded at S7–10, not so in female. Bates's smaller than other two, both sexes with metallic blue face. Male epiproct longer than cerci, in others equal or shorter. Labium and nonmetallic parts of thorax brown to bright salmon, especially obvious on underside. **Habitat and Behavior:** Presumably breeds in forest streams; most often seen hanging up in shrubs inside forest. Clearly most common of the three emeralds from number of sightings, but still few in number. Has been seen circling at forest edges at dusk and occasionally during overcast days in forest roads and light gaps. **Range:** Caribbean slope, 300–1000 m. Nicaragua to Peru and Brazil.

male

female

Cerro Campana Emerald   *Neocordulia campana*

TL 55 mm; HW 35 mm. **Identification:** Much like Bates's Emerald but slightly larger, both sexes with face reddish to brown rather than dark metallic. Male epiproct not extending beyond cerci and genital lobe on S2 extending well beyond hamules (about even in Bates's). **Habitat and Behavior:** Presumably forest streams. Found in CR only once, near Colonia Virgen del Socorro, Alajuela Province, in July 1967. **Range:** Caribbean slope, 850 m. Costa Rica and Panama. Not illustrated.

Elusive Emerald   *Neocordulia griphus*

TL 54 mm; HW 37 mm. **Identification:** Very similar to Bates's Emerald, both with metallic blue face, but male epiproct barely extends beyond cerci or not at all. In addition, cerci gradually increase in diameter to just before tip, while those of Bates's of nearly uniform diameter for most of length, with abrupt swelling just before tip. Genital lobe as in Bates's, not extending beyond hamules and thus differing from Cerro Campana. Appendages of Elusive much like those of Cerro Campana, but face color distinguishes them. **Habitat and Behavior:** Presumably forest streams. Only record from CR a pair collected at Reserva Forestal San Ramón, Alajuela Province, in June 1968. **Range:** Caribbean slope, 1000 m. Guatemala to Costa Rica. Not illustrated.

Skimmers (family Libellulidae)

One hundred one species. These are the familiar, usually brightly colored dragonflies that we see around ponds everywhere in the world. A small subset lives on streams. Most perch prominently and fly out to catch passing insects like a flycatcher. Of the 29 genera that occur in Costa Rica, here are a few distinctive ones, starting with pond species. *Perithemis* are the smallest, males with yellow-orange wings. *Micrathyria* are diverse, from tiny to midsize; all have green eyes, usually a striped thorax, and a black abdomen with prominent pale markings on S7. *Nephepeltia* are very similar, both species tiny; they differ from *Micrathyria* in wing venation.

Erythrodiplax are among the most common dragonflies in open ponds everywhere and are much more varied in color pattern than most genera, from red to blue to black, some with red- or black-patterned wings. *Erythemis* are larger, dashing predators that often catch other odonates; they also vary in color, red and black and blue, with one very large green species. Common large red dragonflies are usually *Orthemis*, with several similar species. Other good-sized red lowland pond species include two *Erythemis*, *Brachymesia furcata*, and *Rhodopygia hinei*. *Libellula herculea* is unmistakable, with pruinose thorax and carmine abdomen. The brilliant scarlet *L. croceipennis*, with amber-tinted wings, occurs at upland streams. Two red *Sympetrum* species are only at high elevations.

Some skimmers of ponds are fliers, spending much of their time on the wing. The most common are *Pantala* and *Tramea*, very widespread and occurring north to the US. If the water surface is covered by floating vegetation, the slightly smaller *Tauriphila* and *Miathyria* fly over mating territories. All of these genera assemble in feeding swarms over open country. The cryptic *Tholymis* is also a flier but comes out at dusk like crepuscular darners.

Brechmorhoga and *Macrothemis* are the common stream skimmers and are also fliers, flying up and down streams and landing on streamside vegetation. They may cruise over clearings in feeding aggregations, sometimes in groups. Most species have a dark abdomen with pale markings on segment 7 much like the clubtails that may be seen on the same streams. Most *Brechmorhoga* are larger than *Macrothemis*. *Dythemis* breed in streams but may also occur on ponds. They look something like the preceding genera but are perchers.

Inside forest, watch for *Uracis*, two of three with black-tipped wings; females have spike-like pseudo-ovipositors. They are often the most common dragonflies along forest trails. *Anatya* and *Cannaphila* are also seen in the same habitat, and immature and female individuals of almost any member of this family might be in a forest clearing. Some of them, including the blue-winged *Zenithoptera*, feed from high in the trees, so looking up is always worthwhile.

Note that any member of this family may be called a skimmer, even if that word is not in the common name.

KING SKIMMERS   genus *Libellula*

Four species. *Libellula* is a temperate-zone genus, with species extending into the tropics, mostly in the highlands. Hercules Skimmer is the only species with a wide range in the tropical lowlands. They are among the largest of skimmers, with bulky thorax and robust abdomen. Most have broader abdomens than tropical king skimmers (*Orthemis*), but females of both genera have flaps on S8 used to pick up water drops and splash eggs onto the adjacent bank, usually just above water level, where they are safe from aquatic predators but close enough that the larvae can reach the water easily.

TL 55 mm; HW 38 mm. **Identification:** Size and robustness, colorful wings, and broad scarlet abdomen distinctive. Male Hercules Skimmer (p. 242) has carmine abdomen, pruinose silvery sides to thorax, and dark wing veins. Females very similar, but stigmas in Neon are orange, in Hercules brown. Neon has reddish wing veins and mostly orange flap on S8; these brown and black, respectively, in Hercules. Similarly colored Brilliant Redskimmer (p. 274) is smaller, with narrower abdomen, and a lowland species. Various red pondhawks and other species are all less robust than king skimmers. **Habitat and Behavior:** Ponds and marshy areas along streams and spring seeps in sun. Males display by elevating abdomen while hovering facing another male or female. **Range:** Pacific slope including Central Valley, to 1450 m. Southwestern US to Colombia.

male

female

241

TL 52 mm; HW 39 mm. **Identification:** Very large and robust species. Males with silver-sided pruinose thorax and intensely red abdomen; females most robust of CR skimmers, abdomen broad to its tip with conspicuous black flaps. See Neon Skimmer (p. 241). **Habitat and Behavior:** Adults seen at ponds, temporary pools, and areas near streams and spring seeps in sun. Often perch high up on tree branch. Known breeding sites have been artificial phytotelmata (cups nailed to trees), so may breed in tree holes. Absence of territorial males from waterside often indicates this, as does sighting of one female ovipositing in a rain barrel. Also breeds in normal habitats, as male seen defending and female ovipositing in puddle on muddy road, another female ovipositing in small forest pond. **Range:** Both slopes, to 1600 m. Eastern Mexico to Argentina.

male

female

TL 48 mm; HW 38 mm. **Identification:** No other regional species of fairly large size with blue-gray coloration and dark eyes and face. Maria's Skimmer (p. 244) most similar but has darker thorax and dark streaks at wing bases. Eastern Pondhawk considerably smaller and with turquoise eyes and green face; no overlap in elevation. Females much like female Maria's but wings without reddish tint; smaller than similarly colored Neon (p. 241) and Hercules Skimmers. **Habitat and Behavior:** Marshy pastures and ponds, usually associated with cloud forest. Quite uncommon and local. **Range:** Both slopes, 1300–1900 m. Highlands of southern Mexico to Panama.

male

TL 50 mm; HW 43 mm. **Identification:** Males with black-tipped blue-gray abdomen, similar to Highland Skimmer (p. 243) but slightly larger, with dark brown instead of blue-gray thorax, and with basal dark stripe in wings. Females even more like Highland Skimmer but with more orange extending along anterior edge of wings. **Habitat and Behavior:** Ponds, temporary pools, and slow pools in small streams, usually associated with cloud forest. Seen mating and ovipositing at deep pool formed by root hole of big fallen tree in swamp forest. **Range:** Both slopes, 1100–1900 m; uncommon and local. **Costa Rican endemic.**

male

male

female

TROPICAL KING SKIMMERS genus *Orthemis*

Seven species. Tropical king skimmers can be divided by abdomen shape into two types. The abdomen is broad and robust in Carmine, Red-tailed, Roseate, Side-striped, and Yellow-lined, slender and appearing relatively longer in Slender and Swamp. Females of all species have prominent flaps on S8. Any of them may breed in small temporary pools, puddles in dirt roads being especially attractive, and the larvae burrow in mud, so aquatic vegetation is not necessary. Females often perch up in trees well away from water, also on utility wires. Copulation is brief and takes place in flight, and then the female oviposits with the male in attendance. Females use their abdominal flap to splash eggs up onto a moist bank rather than dropping them in the water, as most libellulids do. *Orthemis* are among the most common and conspicuous red dragonflies of Costa Rica, although numerous other genera also have red species, including pennants, pondhawks, dragonlets, king skimmers, rock skimmers, scarlet-tails, redskimmers, meadowhawks, gliders, and saddlebags. Red is highly visible against green vegetation, brown soil, and blue water! Generic differences are evident in wing venation, but shape and size will help sort them out in the field. Among red dragonflies, *Orthemis* stand out by not having a trace of color at the wing bases.

Roseate Skimmer *Orthemis ferruginea*

TL 52 mm; HW 41 mm. **Identification:** Males with darker head and eyes than other similar red *Orthemis*; more purplish overall, but robust species all similar enough always to require close scrutiny. Dark markings low on thorax of mature males left over from immature phase are a good character for this species. Wing veins also somewhat paler, more orange-tinted than those of Carmine and Red-tailed Skimmers. Because of paler wing veins, somewhat like male Side-striped Skimmer (p. 249), but latter has red face and eyes like Carmine. Females and immature males distinctive because of more complex thoracic pattern, with pale cross-stripe between two diagonal stripes on side isolating two round spots. **Habitat and Behavior:** Breeds in every type of still water in the open, from pools in streams to pond and lake edges (even in puddles in dirt roads). Most likely species of the genus to occur in entirely open areas without adjacent forest. **Range:** Pacific slope, to 1300 m; few records from northern Caribbean slope, south to Sarapiquí region. Southern US to Costa Rica, Cuba.

male

female

TL 51 mm; HW 39 mm. **Identification:** Mature males most uniformly red of regional *Orthemis*. Red-tailed Skimmer (p. 248) with non-pruinose scarlet abdomen contrasting with darker thorax. Roseate Skimmer (p. 245) more pinkish-purplish, with darker eyes and face. Mature male Side-striped (p. 249) most similar in redness but with whitish edges to face and reddish wing veins; so far known only from southern Pacific region. Female Carmine with thorax only lightly marked, un-like strong diagonal stripes of female Red-tailed and Roseate and horizontal stripe of Side-striped. When female Carmine does show stripes, pattern is more subdued version of Red-tailed stripes. Carmine usually shows black abdominal flaps, Red-tailed brown, but some females may be in-distinguishable. **Habitat and Behavior:** Breeds in temporary and permanent ponds, pools, road puddles, ditches, and spring seeps in sun; more commonly associated with forest than Roseate. Territorial interactions can involve up to four males in rapid flight, twisting and turning in unison and ranging from high in the air to skimming the water, one individual returning to territory but then flight beginning again in a few seconds. **Range:** Both slopes, to 1400 m. Texas to Argentina.

male

female

TL 50 mm; HW 39 mm. **Identification:** Males readily distinguished by strong contrast between reddish-purple pruinose thorax and bright scarlet non-pruinose abdomen. Females distinguished from Carmine Skimmer (p. 247) by almost always more heavily striped thorax and brownish abdominal flaps, from others in robust group by pattern on thorax. **Habitat and Behavior:** Breeds in ponds in or out of forest. **Range:** Both slopes, to 400 m. Guatemala to Bolivia and Brazil.

male

female

TL 50 mm; HW 39 mm. **Identification:** Mature males redder than Roseate Skimmer (p. 245), lacking purplish to pink cast. More like Carmine Skimmer (p. 247) but with reddish wing veins (vs. black) and a narrow light stripe often evident on lower edge of thorax. Sides of face whitish in Side-striped, while red covers entire face in Carmine. Roseate has more purplish face, sometimes paler on sides. Pale horizontal stripe on each side of thorax makes females and young males unmistakable when compared with Carmine and Roseate Skimmers, which have diagonal stripes following thoracic divisions. **Habitat and Behavior:** Appears to be a pond breeder, in both forested and deforested landscapes. Commonly seen well away from water, as are some other *Orthemis*. A sleeping aggregation of this species, with numerous individuals hanging vertically and close together, was found once on the Osa Peninsula. Presumably this is normal behavior, but it has been very rarely seen in New World dragonflies. **Range:** Pacific slope, north at least to Quepos, to 1200 m. Costa Rica to Argentina.

male

female

TL 45 mm; HW 37 mm. **Identification:** Contrast between dark brown thorax (with five vivid cream-yellow stripes) and bright red to red-brown abdomen unique among CR skimmers. Stripes may not be visible from directly above. Females and immature males of others, especially Red-tailed Skimmer (p. 248), have thoracic stripes, but not as contrasty, and with abdomen light brown. Slightly smaller than other robust *Orthemis*. **Habitat and Behavior:** Apparently largely restricted to swamps (forested wetlands). Poorly known. **Range:** Pacific slope, to 300 m; north to Quepos. Guatemala to Bolivia and Brazil.

TL 45 mm; HW 36 mm. **Identification:** Conspicuously striped black thorax compared with overall reddish or brown thorax of even more attenuate Slender Skimmer (p. 252). Note that immature male Slender has striped thorax, but its dark sections are brown, paler than metallic blackish of Swamp. Slender has much more red on abdomen, with no hint of pale basal stripes or black lateral stripes. Females more similar, but Slender has more pale than dark color on thorax, Swamp more dark than pale. Note that front of thorax in Swamp is nearly black, while Slender has bold pale frontal stripes. Few skimmers in CR show such a slender abdomen as these two species, but compare with slender pondhawks (pp. 275–283). **Habitat and Behavior:** Breeds in forest ponds and swamps, rarely seen outside forest. Males alternately hover over water and perch 1 to 2 m up. **Range:** Pacific slope, north to Quepos, and entire Caribbean slope, to 100 m. Costa Rica to Argentina.

male

female

251

TL 49 mm; HW 35 mm. **Identification:** See somewhat similar but less slender Swamp Skimmer (p. 251) for color differences. Male Slender could be mistaken for male Red Pondhawk (p. 278), equally long and slender, but pondhawk has expanded abdomen base not present in skimmer, also no black at abdomen tip. **Habitat and Behavior:** Breeds in forest ponds and pools in forest streams, also in open marshes with dense vegetation; not tied to forest like Swamp Skimmer. Males may also set up territories over very small rain puddles on forest trails. **Range:** Both slopes, to 900 m. Central Mexico to Venezuela.

male

female

CONVICT SKIMMERS genus *Cannaphila*

Three species. These medium-sized skimmers of forested swamps and streams all have a white face and a strongly striped thorax, and all change appearance with maturity. They commonly perch lengthwise along a branch or stem, usually with the abdomen pointed downward, rather than across or at the tip as in most skimmers. Females have flaps on S8 like those of tropical king skimmers, to which they are related.

Gray-waisted Skimmer *Cannaphila insularis*

TL 42 mm; HW 31 mm. **Identification:** No other regional dragonfly black with gray base of abdomen. Easily distinguished from Blue-tailed Skimmer (p. 256), which has a pruinose blue abdomen. Females more similar, but Blue-tailed always has light stripe between wings and down front of thorax, conspicuous and about twice as broad as similar stripe in Gray-waisted. Both sexes of Gray-waisted vary greatly, with different combinations of yellow, orange, gray, and black, such that assigning an age to some individuals difficult, although in general they become darker and less patterned with age. Orange immatures darken with maturity and develop metallic blue on top of frons. Old females look like males but usually duller and darker gray. Occasional females have dark wing tips or even completely brown wings. Caribbean-side individuals smaller by a few millimeters on average. See quite similar Morton's Skimmer (p. 255). **Habitat and Behavior:** Isolated ponds and still pools in streams in forested landscape. Males hold territories over swamp pools and other small bodies of water, including tiny streams. **Range:** Both slopes, to 750 m. Texas to Panama, Greater Antilles.

male

old female

female

254

TL 36 mm; HW 28 mm. **Identification:** Very similar to Gray-waisted Skimmer (p. 253), only slightly smaller and much more restricted in range and habitat. Positive identification by distinctly narrower base of hindwings, readily seen in photos that show wings well. Anal loop typically short, with 6–9 cells in this species, 9–12 cells in Gray-waisted. Forewing triangle uncrossed in this species, crossed in Gray-waisted. Mature males pruinose between wings and on base of abdomen, with much red-orange on middle segments. Both sexes with much color on top of abdomen into maturity, unlike Gray-waisted. Posterior hamule more obvious than that of Gray-waisted in lateral view, hook-shaped. **Habitat and Behavior:** Has been seen at forested ponds and swamps and near sources of small streams, but larval habitat and basic natural history not known. Immatures in grassy areas at forest edge. **Range:** Caribbean slope, to 600 m. Costa Rica to western Ecuador.

male

female

TL 40 mm; HW 31 mm. **Identification:** Rarely occurs with Gray-waisted Skimmer (p. 253), distinguished by more and bluer pruinosity on abdomen and persistence of thoracic stripes throughout life. Conspicuous pale median stripe on front and top of thorax diagnostic. Anal loop boot-shaped, straight and sac-shaped in other two convict skimmers. Because of thoracic stripes, mature males not likely to be confused with other pruinose blue species, including Eastern Pondhawk, Montane and Andagoya Dragonlets, and Highland Skimmer. **Habitat and Behavior:** Typically associated with marshes along streams, but also at natural and artificial ponds and pools. More likely than Gray-waisted to perch with wings depressed. **Range:** Both slopes, to 1600 m but mostly above 600 m. Highlands of eastern Mexico to Argentina.

male

female

female

AMBERWINGS   genus *Perithemis*

Four species. Amberwings are the smallest of Costa Rican dragonflies. Males are distinctive because of yellow to orange wings; female wings vary from transparent to amber, usually with complex dark patterns. Males fly and hover low over the water, usually in the sun, in both open and forested habitats. They also spend much time examining potential oviposition sites, even dipping the abdomen as if ovipositing, and then lead females to them and mate there, usually hovering very near to drive off other males. Females often perch in weedy areas away from water, and both sexes could be considered wasp mimics, with spindle-shaped abdomens and colored wings. Usually no more than two species are present at a given site.

Slough Amberwing   *Perithemis domitia*

TL 24 mm; HW 18 mm. **Identification:** Males distinguished from other amberwings by striped thorax, dark legs (tibiae usually paler), and straight abdominal stripes; abdomen looks black above and orange below. Females show these characters and further distinguished by orange-based wings, with brown spots or bar at nodus. Minority of both sexes sometimes have additional brown spots in wings. Rare female variant has orange on wings paler, almost colorless. **Habitat and Behavior:** Breeds in slow streams and sloughs with forested banks, also at forested natural or artificial ponds. **Range:** Both slopes, to 800 m. Arizona and Texas to Ecuador and Brazil, Greater Antilles.

male

female

TL 22 mm; HW 18 mm. **Identification:** Patterns on thorax and abdomen distinctive. Thorax brown in front with paired yellowish oval patches, yellowish on sides with trace of brown stripe. Abdomen with dark dorsolateral stripes converging toward rear on each segment to form triangles. Females with variable wing patterns, some with nothing more than yellowish wash on front edge of clear wings, but none looks just like female Slough. Only amberwing with forewing subtriangle uncrossed; others with single crossvein. Thorax looks somewhat striped, but pale legs distinguish Eastern from Slough Amberwing, with more distinctly striped thorax and dark legs. (Pond Amberwing, *Perithemis mooma*, was long considered a separate species but is now combined with Eastern Amberwing.) **Habitat and Behavior:** Breeds at open ponds and lakes, including seasonal ones, also at times at pools in sunny streams. **Range:** Both slopes, to 1000 m. Southern Canada and eastern United States to Argentina.

male

female

female

TL 22 mm; HW 17 mm. **Identification:** Sexes colored identically, uniformly pale thorax and legs distinguishing them from other amberwings. Those with more extensive abdominal markings could be confused with Eastern (p. 259) or Slough (p. 258) Amberwings if thorax not seen clearly. In addition, rare females of other two species may have largely pale, often orange-tinged wings, and these could be mistaken not only for female Golden but males of any species. Good look at thoracic and abdominal patterns should allow identification. Golden has slightly broader wings than Slough and Eastern. **Habitat and Behavior:** Breeds at forest pools and swamps with sun penetration. **Range:** Both slopes, to 300 m. Guatemala to Bolivia and Brazil.

male

female

261

TL 22 mm; HW 18 mm. **Identification:** Unmistakable in both sexes by orange wings with narrow to wide brown bands across them, either mid-wing or mid-wing and tip. Abdomen with more black on top than other amberwings, longitudinally striped as in Slough (p. 258) rather than with diagonal markings as in Eastern Amberwing (p. 259). Stigmas black; in all other species stigmas are orange to red. Males oddly polymorphic, a small percentage with extensive dark wing tips and limited basal markings. Female wings colored somewhat between the two male types. Female Slough and Eastern Amberwings have dark wing markings but not on entirely orange wings. Occurs most commonly with Golden Amberwing. **Habitat and Behavior:** Breeds at forested ponds, generally perching well above water. Males perform 'dance' in which one male hovers and another flies close in front of it in semicircular path back and forth. Males court females using a different "dance," in which both individuals fly slowly upward and then increasingly rapidly downward in J-shaped curves and then repeat the dance as many as six times. **Range:** Northern Caribbean lowlands, to 100 m; known from Boca Tapada to La Selva. Also reported from southern Nicaragua.

male

rare male type

female

263

TROPICAL PENNANTS genus *Brachymesia*

Two species. These two dissimilar-looking species are united by wing venation and larval morphology, and females of both have rather long cerci. They prefer ponds and lakes larger than those inhabited by many other skimmers. Common name of pennants refers to behavior of perching at tops of stems, flaglike.

Red-tailed Pennant *Brachymesia furcata*

TL 43 mm; HW 34 mm. **Identification:** Should be compared with many other skimmers that are almost entirely red. Because of minimal wing markings, not to be mistaken for saddlebags or gliders; more like tropical king skimmers (*Orthemis*, p. 245), but smaller, with differently shaped abdomen. Note upward curved cerci and black dorsal markings at abdomen tip. Very few plain brown dragonflies like females (some develop red abdomen like males), none of them with cerci as long as in this species. Brilliant Redskimmer (p. 274) larger with redder eyes and face and unmarked abdomen. Also check female streamskimmers, pondhawks, scarlet-tail, and redskimmer. **Habitat and Behavior:** Ponds in open areas, with scattered emergent and floating aquatic vegetation, and sometimes inundated shrubs. More likely at lakes and larger ponds. Males perch on tips of vegetation and chase any other red dragonflies that appear, including larger species. Females often perch in shrubs near water. **Range:** Both slopes and Central Valley, to 1400 m. Southern US to Argentina, West Indies.

male

female

264

TL 45 mm; HW 36 mm. **Identification:** Overall brown coloration and long, slender abdomen with black dorsal stripe unique. Abdomen looks swollen at base, much more slender and parallel-sided from S4–10. Wings heavily suffused with golden-brown, darker toward front, and at maturity may be quite dark. Fully colored wings not shared with any species even vaguely similar in appearance. Females with wings not so heavily washed with brown, and small orange spot at base of hindwing also distinctive, as are long cerci. **Habitat and Behavior:** Like Red-tailed, a species of larger bodies of water. Males perch at tips of stems and inflorescences of tall sedges and grasses or tips of dead twigs over water at shores of open ponds and lakes. Females perch like males but away from water; both sexes at times as high as utility wires. Wings usually held up at shallow angle, increasingly upward with wind speed and probably a source of stability. Seen capturing large termites in late afternoon, chewing off wings and eating remainder. **Range:** Both slopes, to 600 m. Southern US to Argentina, West Indies.

male

female

TL 38 mm; HW 27 mm. **Identification:** This species is small, dark, and slender, with shiny black abdomen in both sexes and small dark spot at base of hindwings. No other species much like it, although male Seaside Dragonlet (p. 294) superficially similar; latter does not perch like pennant and is not shiny. Females duller than males, with more pale markings on thorax and yellow line along outer edges of abdomen to variable distance, in some to S7. **Habitat and Behavior:** Ponds, more common at larger ones. Males perch at tips of plant stems and inflorescences, usually at outer edge of vegetation, and fly out over open water, sometimes in flight for lengthy periods and with some hovering. Females perch well up in trees both close to and away from water. **Range:** Caribbean slope, to 500 m; so far known only from Los Chiles, Boca Tapada, and La Marina in Alajuela Province and Puerto Viejo de Sarapiqui in Heredia Province. Texas and Florida to Venezuela and Greater Antilles.

male

female

TL 26 mm; HW 18 mm. **Identification:** One of smallest dragonflies in CR, easily recognized by bright pattern of pale markings on thorax and abdomen base, including vertical stripe on S2 and small, almost joined spots on S7; mature individuals have blue eyes. Males with swollen abdomen base, slender beyond that and then slightly clubbed. Broad amber suffusion at base of wings in both sexes unique among similar species. Pale markings more extensive than in somewhat similar Little Swamp Dasher (p. 315) and dryads (p. 317), also no trace of pruinosity. Also called Bluc-eyed Fairy. **Habitat and Behavior:** Territorial males at sand-bottomed forest streams, where females oviposit without tapping water. Also feed at forest edge. **Range:** Northern Caribbean slope, to 200 m; recorded so far only at Reserva Biológica Tirimbina in Heredia Province. Costa Rica to Peru and Brazil.

immature male

immature female

TL 38 mm; HW 29 mm. **Identification:** Legs longer than those of other red skimmers. Males with prominent dark brown hindwing patches, dark blue and somewhat pruinose thorax, and bright scarlet red abdomen with S1-2 pruinose above. Flame-tailed Pondhawk (p. 280) only dragonfly in CR like male, but pondhawk has obvious dark base to abdomen, including S3, and no hindwing patches. Female scarlet-tail unique in having dark brown thorax and reddish-brown abdomen with large hindwing patches. Female saddlebags (pp. 352–355) somewhat similarly colored but larger and much more often in flight; also with more slender abdomen with black markings at tip. **Habitat and Behavior:** Typically large ponds, where males perch elevated on long legs, on leaves and stems well out over open water. Raises abdomen to obelisk on hot middays. Females perch at edges of forest clearings, often up in trees. **Range:** Northern Caribbean slope, to 100 m. Texas to Costa Rica.

male

female

TL 42 mm; HW 33 mm. **Identification:** Note wing markings, extensive in males and filigree-like in females. Oddly patterned abdomen and finely lined eyes also distinctive. Note some similarity to both Black-winged (p. 297) and Band-winged (p. 295) Dragonlets, neither of which is likely to be perched on rocks or ground at streams. **Habitat and Behavior:** Males perch on rocks and fly up and down streams, usually rocky ones. Both sexes commonly perch on ground. **Range:** Two records on Pacific slope, from a century ago near Turrúcares and in 2014 near Ostional, on Nicoya Peninsula. Southwestern US to Costa Rica.

male

female

TL 43 mm; HW 25 mm. **Identification:** Mature males and females only tropical forest dragonfly that combine blue eyes and mostly blue thorax. Females apparently polymorphic, with blue or tan thorax. Immatures and tan females identifiable by unique pattern on thorax, complex but darkest on front, and spotted abdomen. In males, white appendages with upcurved cerci and slender abdomen are distinctive. Other smallish skimmers with spotted abdomens include tropical dashers (*Micrathyria*, p. 301), which have green eyes, striped or otherwise differently patterned thorax, and dark appendages. Blue-eye has pale spot on S6 usually larger than on S7; spot on S7 largest in dashers. **Habitat and Behavior:** Usually seen at swamp ponds and still pools in forested streams; females and immatures along forest trails, often in tall herbaceous vegetation. Spend most time in shade but don't entirely shun sunlight. Probably long-lived, as individuals with algae on wings and even body often seen. **Range:** Lowlands of both slopes, to 400 m. Central Mexico to Bolivia and Brazil.

male

female

WOODSKIMMERS genus *Uracis*

Three species. These skimmers are common forest dragonflies, usually seen along trails under the canopy or at the forest edge. They breed in seasonal puddles, where females use specialized subgenital plates called pseudo-ovipositors to jab eggs into the mud in low areas, even before they flood. Two of our three species have prominent dark wing tips. Thoracic pattern of immature woodskimmers unique to this genus, with fine dark horizontal lines on front of thorax producing almost wood-grain effect.

Large Woodskimmer *Uracis fastigiata*

TL 38 mm; HW 32 mm. **Identification:** Good-sized dragonfly with outer third or more of wings black; males with dark thorax and blue-gray abdomen, females dark gray. Females with very long, pointed subgenital plate projecting well beyond abdomen tip. Smaller Tropical Woodskimmer easily distinguished by much less black at wing tips and shorter subgenital plate in females. At a glance, especially in flight, possible confusion with male or andromorph female Band-winged (p. 295) or Black-winged (p. 297) Dragonlets because of so much black in wings. This species seems especially likely to accumulate algae on wings and body, perhaps indicating very old individuals surviving through dry season. **Habitat and Behavior:** Forest edge and forest understory, probably breeding in rainpools in wet season and spending dry season (with immature coloration) in reproductive diapause. **Range:** Both slopes, to 1200 m. Guatemala to Bolivia and Brazil.

male

female

TL 35 mm; HW 26 mm. **Identification:** Smaller, entirely pruinose-gray woodskimmer with crisply defined, but limited, black wing tips, much less extensive than in Large Woodskimmer. Female subgenital plate much shorter, projecting only slightly beyond abdomen tip. Only other skimmers with sharply defined black wing tips are females of Gray-waisted Skimmer, Blue-eyed Setwing, Delta-spotted and Jade-striped Sylphs, and Swamp Dasher, all differently colored otherwise. **Habitat and Behavior:** Clearings at forest edge and within forest understory, breeding in rainpools in wet season. Often abundant during dry season in lowland riparian forest on Pacific slope, where immature in coloration and in diapause (non-reproductive) until beginning of next wet season. **Range:** Both slopes, to 900 m. Eastern Mexico to Argentina.

male

female

TL 37 mm; HW 31 mm. **Identification:** Easy to distinguish from other woodskimmers by clear wing tips; female subgenital plate intermediate between those of Large and Tropical. Distinguished from all other skimmers by front of thorax either finely barred or entirely dark, contrasting with black-and-pale-tan patterned sides. Black S7–10 in both sexes contrasts with both pruinose blue-gray (mature) or orange (immature) abdomen base and with snow-white cerci. **Habitat and Behavior:** Very small streams, marshes, and swamps in forest. Male observed guarding and patrolling a small pool in a spring-fed rivulet within forest. Female accompanied by male oviposited at midday in muddy gravel 1.5 m from forest stream. Many tenerals seen in forest edge bordering a marshy spring seep. May spend dry season as immature adult, as in other woodskimmers. **Range:** Caribbean slope, to 1000 m. Belize to Costa Rica.

male

female

TL 48 mm; HW 38 mm. **Identification:** Large and robust species and the brightest red of the red CR skimmers, glowing scarlet in sun. Males of most other red skimmers have noticeable distinguishing characters: Neon Skimmer (p. 241) larger and with orange wings; *Orthemis* species (p. 245) larger and duller red (but Red-tailed Skimmer with similarly bright abdomen); Red-tailed Pennant (p. 264) and Red Pondhawk (p. 278) not so red and both with very different shape; and Cardinal and Talamanca Meadowhawks with slender abdomen and yellow-spotted thorax, also smaller and at high elevation. Most similar are equally bright Claret Pondhawk (p. 279) and Scarlet Dragonlet (p. 284), smaller and with more slender abdomens and prominent dark hindwing patches. Contrast between brown front and red sides of thorax in some male redskimmers also distinctive. Females with no markings at all, whether all brown or with brightly contrasting red abdomen, and no wing markings other than orange tint at base. Both sexes distinctive from above, with abdomen expanding very slightly toward rear and widest just before tip, unique to this species. **Habitat and Behavior:** Breeding habitat not well known; males seen perched over water at forest ponds, pools in small streams, and roadside ditches with flowing water, often with much vegetation. Tends to be wary and difficult to approach closely. Males hover just above water surface with abdomen inclined somewhat upward, behavior not seen in other red dragonflies. Females perch along forest trails, typically on branches or trunks in near vertical position. Oviposit by tapping water regularly and moving between taps. **Range:** Both slopes, to 1000 m. Guatemala to western Ecuador.

PONDHAWKS   genus *Erythemis*

Eight species. These medium to large skimmers are among the most common in their family throughout the New World tropics. Like species of *Orthemis*, they exhibit two different shapes, four species of average abdomen length and width and four species with especially long, narrow abdomens with swollen bases. Great Pondhawk, the largest, is the longest member of its family in Costa Rica. This shape dichotomy begs explanation, but color and shape can be used in combination for identification. Both "short-tailed" and "long-tailed" species include a black, a red, and a green species. Eastern Pondhawk, the short green one, has males that turn blue when mature.

At least some pondhawks regularly take large prey such as butterflies and other dragonflies. All species have three prominent long spurs on the underside of the distal end of the hind femur, probably an adaptation for capturing such robust prey; dragonlets and most other skimmers have only small spines, although they are relatively long in redskimmers and leafsitters. Most pondhawks prefer heavily vegetated wetlands, even a complete cover of floating vegetation, where they perch for long intervals and periodically fly low over their territories. Females of all forage well away from water, often along forest trails. Perching on the ground is particularly noticeable in this genus. Females of some species (Black, Red, Claret, Flame-tailed, and Pin-tailed) have a pointed subgenital plate projecting below the abdomen at S8 and visible from the side. This distinctive structure is also found in some dragonlets, a closely related group. Copulation is brief, first in flight and then perched; after separation, both sexes remain perched for a short spell before the female commences oviposition. Males usually guard egg-laying females.

Black Pondhawk   *Erythemis attala*

TL 42 mm; HW 35 mm. **Identification:** Solid black males and mature females distinctive. Rather similar to Pin-tailed Pondhawk (p. 277), but with abdomen relatively shorter and without such extreme narrowing beyond base; also note hindwing spots distinctly larger. Pin-tailed tends to occur in more open environments. Female and immature Black distinctive from other dark species because of four pairs of large yellow spots on abdomen and pale appendages. Spots can persist well into maturity but disappear in most individuals. (Mature females and some males from Panama south have two bright yellow bands on abdomen at S4 and S7 instead of multiple spots, could represent a different species and should be watched for in CR.) **Habitat and Behavior:** Males at forested ponds with much floating vegetation such as water lettuce and floating fern, often covering surface. Females and immatures in sunny spots in forest, usually perching above ground. Large hindwing spots make males quite conspicuous in flight. **Range:** Both slopes, to 700 m. Texas to Argentina, Greater Antilles.

male

female

TL 45 mm; HW 31 mm. **Identification:** Characterized by slender abdomen with prominently enlarged base in side view; brown with black bands in immature and females, entirely black in mature males (some males keep pale bands to maturity). Abdomen same length as hindwing and slender in comparison with that of Black Pondhawk (p. 275); hindwing spot smaller. Similar in shape to female Red Pondhawk (p. 278) but latter browner, without paler thoracic front, and abdomen much less conspicuously banded. Female Slender Skimmer (p. 252) with unbanded abdomen, never perches on ground. **Habitat and Behavior:** Breeds at open ponds and marshes, not tied to forest as are some other pondhawks. Males almost always guard egg-laying females; one seen ovipositing on floating algal mats, tapped 1–10 times, each at different spots. **Range:** Both slopes, to 600 m. Texas to Argentina, also Florida and Greater Antilles.

male

female

TL 47 mm: HW 33 mm. **Identification:** Long-bodied red dragonfly with darker thorax, most similar to smaller and shorter-bodied Claret Pondhawk except for shape. Could be mistaken for other slender red species, Slender Skimmer (p. 252) for example, but males of that species have black abdomen tip and appendages. Both sexes may have dark band at end of each middle abdominal segment. **Habitat and Behavior:** Males at ponds and marshes, usually associated with forest. Females forage along forest trails. **Range:** Both slopes, to 300 m. Northern Mexico to Peru and Brazil, Greater Antilles.

male

female

TL 36 mm; HW 29 mm. **Identification:** Both sexes distinguished from larger, more slender Red Pondhawk by shorter, broader abdomen and larger hindwing spots (note hindwing spots on Pintailed and Black Pondhawks). Male Scarlet Dragonlet (p. 284) very similar to Claret but smaller and somewhat brighter red all over. Dark hindwing spots of dragonlet patterned by paler veins, not so in pondhawk. Claret Pondhawk has pale stigmas, dragonlet dark. Female Scarlet Dragonlet has no hint of dark borders on abdominal segments and has distinctly shorter subgenital plate, easily seen in side view. Note other bright scarlet males such as Red-tailed Skimmer (p. 248) and Brilliant Redskimmer (p. 274), neither of which has prominent hindwing spot. **Habitat and Behavior:** Forested ponds with much floating vegetation. Immatures and females in forest clearings. **Range:** Pacific slope, to 100 m; likely to occur on Caribbean slope but no records. Texas to Argentina.

male

female

TL 42 mm; HW 28 mm. **Identification:** Short, stout pondhawk; male thorax pruinose blue-black and abdomen bright red with basal segments black. Nothing else like it except male Mexican Scarlet-tail (p. 268), which has almost entire abdomen red. Scarlet-tail has very long legs that hold it well above substrate in comparison with the short-legged pondhawk. Abdomen of male Flame-tailed becomes red before thorax becomes pruinose, so look for occasional individuals with tan and brown thorax and red abdomen. Females could be mistaken for female Black Pondhawk (p. 275) or dragonlets, but contrasty thoracic pattern distinctive. Females of smaller Little Pondhawk (p. 283) and much smaller Red-faced (p. 286) and Montane (p. 288) Dragonlets have similar pale-fronted thoracic pattern but show dark dorsal stripe on abdomen; Flame-tailed markings are obscure and dorsolateral on each segment of broader abdomen. **Habitat and Behavior:** Breeds at ponds and marshes, typically with dense grasses. **Range:** Both slopes, to 300 m. Northern Mexico to Argentina.

male

female

TL 41 mm; HW 32 mm. **Identification:** No other lowland CR dragonfly is all blue with white cerci but note Highland (p. 243) and Maria's (p. 244) Skimmers, two much larger brown-eyed, dark-faced species restricted to highlands. Green immature and female Eastern could be mistaken only for Great Pondhawk, considerably larger and with much narrower, swollen-based abdomen. Eastern perches commonly on ground and flat surfaces, Great only rarely. Note that males may be seen in transition between green immature and blue mature coloration. **Habitat and Behavior:** Males at open ponds and marshes, often perching low, flat on leaves or on ground. Among the most abundant dragonflies in the US, much less common in tropics. **Range:** Northern Pacific slope, to 100 m; so far known from few localities, no known resident populations. Southeastern Canada to Costa Rica, Greater Antilles.

male

female

TL 58 mm; HW 40 mm. **Identification:** Very long green skimmer with dark-banded abdomen. Sexes look alike, females distinguished by cream-colored appendages shorter than those of males. Females and immature males of rare Eastern Pondhawk much smaller; also note shorter abdomen with whitish and dark crossbands that are pointed in front. There are no other green skimmers, but could be confused with darners with green thorax (Amazon, Blue-spotted Comet, Blue-faced, and Mangrove), all of which also fly back and forth over wetlands but hang up when perched and do not have green-banded abdomen. **Habitat and Behavior:** Males patrol small to medium bodies of water, including quiet parts of slow streams. Both sexes common in sunny and shady areas within forest, sometimes far from water. Commonly takes large prey such as butterflies and other dragonflies. One of widest ranging and most familiar of neotropical dragonflies, although not as abundant as many smaller species. **Range:** Both slopes, to 400 m; also Monteverde migration trap, 1450 m. Southern US to Argentina.

male

female

TL 35 mm; HW 25 mm. **Identification:** Males slender and black with gray pruinosity on much of abdomen; dark basal hindwing spots small. Easily mistaken for dragonlet and colored much like Andagoya Dragonlet (p. 289), which is much smaller (two-thirds the length and half the bulk); size alone is distinctive. Andagoya hindwing spot smaller and usually whitish-bordered; cerci whitish. Seaside Dragonlet (p. 294) also all dark, closer in size to pondhawk, but with no hindwing spot and dark cerci; also brackish or marine habitats and equally rare. Brownish thorax of female Little strikingly marked with pale stripe on front and between wings, pale area narrower than in brown females of Flame-tailed Pondhawk (p. 280) and Red-faced Dragonlet (p. 286). Middle segments of abdomen pale with black dorsal stripe, not like any other pondhawk or dragonlet. Blue-gray eyes of female coupled with brown body also unique. Also called Blue Pondhawk. **Habitat and Behavior:** Open and marshy ponds. **Range:** Caribbean slope, to 100 m; known at present only from Boca Tapada. Costa Rica to Argentina.

male

female

DRAGONLETS genus *Erythrodiplax*

Eleven species. This is a diverse group of small to medium-sized skimmers, among the most frequently seen dragonflies in the New World tropics. They are related to pondhawks but are usually much more common, probably because their prey consists of tiny insects that are abundant. Several species typify the breeding strategy of many odonates that breed in seasonal wetlands in the tropics, spending the dry season as adults in sheltered woodland in immature color and then dispersing widely and sometimes in great numbers as the first rains flood the landscape each year.

Two species (Band-winged and Black-winged) are large and black when mature, with black wing markings that distinguish males. Five smaller species (Montane, Andagoya, Red-faced, Chalk-marked, and Coffee Bean) have males with black-tipped pruinose blue abdomens and a variety of markings at the hindwing base. Three species (Scarlet, Red-mantled, and Red-faced) are mostly reddish with some darker coloration at wing bases. Note that Red-faced appears in two categories because of geographic variation. Finally, there are two very rare species, one (Canopy) in forest canopy and the other (Seaside) probably at brackish locations on the Caribbean coast. Females of all but Canopy have a subgenital plate projecting below the abdomen and visible from the side.

Scarlet Dragonlet *Erythrodiplax castanea*

TL 33 mm; HW 25 mm. **Identification:** Small bright red dragonlet. Distinguished from quite similar Claret Pondhawk (p. 279) by smaller size, prominent paler veins in hindwing spot, and absence of three large spines on hind femur. Dragonlet has dark stigmas, pale in Claret Pondhawk. Females difficult to identify because so plain, but lack of abdominal markings distinctive compared with other dragonlets. Also called Leaf-hugging Dragonlet. **Habitat and Behavior:** Shallow, marshy ponds associated with forest. Quite local in comparison with most other lowland dragonlets. **Range:** Caribbean slope, to 700 m; known from relatively few localities. Guatemala to Argentina.

male

female

TL 30 mm; HW 24 mm. **Identification:** On Caribbean slope and in Guanacaste (northern Pacific slope), abdomen of males pruinose blue with reddish base and black tip; in Osa Peninsula (southern Pacific slope), abdomen entirely red (nonpruinose); from northern Pacific to southern Pacific, note variations in abdomen color, with all possible combinations of blue and red. Apparently mature red males occasionally seen on Caribbean slope. Red males much smaller and with less wing color than Red-mantled Dragonlet (p. 291). Blue males much like Montane (p. 288) and Coffee Bean (p. 292) Dragonlets except for reddish face and thorax. Females distinguished from Andagoya (p. 289) and Chalk-marked (p. 290) Dragonlets by pale front of thorax and nonmetallic face. Females and immatures much like Montane Dragonlet, which is larger and usually occurs at higher elevation, but with some overlap from 1000 to 1300 m, and the two have been seen at the same site (Tapantí). **Habitat and Behavior:** Marshy areas of ponds, lakes, and slow streams in full sun; often abundant. Males perch in dense emergent vegetation. Females oviposit in marsh vegetation, usually with male in attendance. Immatures can be abundant at and near water during emergence season. **Range:** Both slopes, to 1300 m; in migration to Monteverde, 1450 m. Texas to Argentina, Lesser Antilles.

male

female

southern Pacific male

TL 35 mm; HW 25 mm. **Identification:** Small dragonfly, males with dark reddish-brown thorax and abdomen base, rest of abdomen light blue with black tip, top of frons metallic blue. Male in blue populations of Red-faced Dragonlet (p. 286) quite similar except for reddish head, with no hint of metallic color, and paler reddish thorax and abdomen base. Younger individuals of both sexes show dark humeral stripe between front and sides of thorax and are difficult to distinguish from some Red-faced, although slightly larger and with abdominal markings more obscure. Look for mature males to be certain, but note that older Montane females become dark and pruinose like males. Also very similar to Coffee Bean Dragonlet (p. 292) but no overlap in elevation. **Habitat and Behavior:** Males fairly common in their upland marshy habitat, at both isolated ponds and pools in slow streams. **Range:** Central highlands, 800–1500 m. Highlands of southern Mexico to Ecuador and Venezuela.

male

female

TL 27 mm; HW 20 mm. **Identification:** Small blue-gray dragonfly, males distinguished from Chalk-marked Dragonlet (p. 290) by usually smaller size and smaller, dark hindwing spot with narrower pale border. However, some Chalk-marked smaller and less heavily marked, looking exactly like Andagoya and must be distinguished by structure of hamules. In Chalk-marked, genital lobe pointed and projects downward; in Andagoya, rounded and directed toward rear. Females and immatures most like Red-faced (p. 286) and Montane (p. 288) in abdominal pattern, distinguished by thorax entirely dark in front instead of pale bordered by dark humeral stripes as in other two. Metallic top of face another distinction from Red-faced. See also Coffee Bean Dragonlet (p. 292) and Little Pondhawk (p. 283). **Habitat and Behavior:** Breeds in grassy, boggy marshes. Often with Red-faced Dragonlet. Immatures with dark reddish thorax can be very common at and near breeding sites. **Range:** Caribbean and southern Pacific slopes, to 300 m. Costa Rica to Ecuador.

male

female

TL 33 mm; HW 23 mm. **Identification:** Another mostly pruinose blue dragonlet, males most similar to male Andagoya Dragonlet but larger, with large black hindwing spot with conspicuous chalky-white border (just a bit of white outside smaller black spot in Andagoya). Some male Chalk-marked have extensive black on middle abdominal segments, only pruinose blue species like that. Younger females and immatures identical to female and immature Red-mantled in color pattern, but metallic top of face distinctive (but in some that effect barely present), also generally less color at wing base in Chalk-marked. See also Coffee Bean Dragonlet (p. 292). **Habitat and Behavior:** Open ponds and marshy wetlands along streams, usually bordered by forest. Sometimes, large numbers of both sexes and all ages at forest edge near breeding wetlands. Males display to each other with abdomen raised and wing spots prominent. **Range:** Both slopes, to 300 m. Costa Rica to Peru and Brazil.

male

female

TL 34 mm; HW 24 mm. **Identification:** Males mostly red with large basal wing markings, extending halfway to nodus in hindwings. Larger than Red-faced Dragonlet (p. 286); often occur together. Duller, darker red than Scarlet Dragonlet (p. 284), with more black on abdomen. Female and immature abdominal pattern like that of Chalk-marked, distinguishable by typically more reddish color at wing bases, although some reddish present in small percentage of Chalk-marked. Metallic color on face, if present, distinctive for Chalk-marked and absent in Red-mantled. Females darken with age, pale areas on abdomen may become quite red. Female wings rarely dark-tipped. Subgenital plate in female Red-mantled straight to slightly convex on rear margin, in Chalk-marked with tip slightly bent backwards, so rear margin slightly concave. **Habitat and Behavior:** Marshes and slow marshy stretches of streams. Can be very common at and near wetlands. Males display to each other with abdomen raised and wing spots prominent, may engage in long chases over water. Dense sleeping aggregation of mixed Red-mantled and Black-winged once found along forest trail at night. **Range:** Both slopes, to 600 m. Northern Mexico to Ecuador and French Guiana, also West Indies.

male

female

TL 32 mm; HW 24 mm. **Identification:** Mature males superficially like Andagoya and Chalk-marked Dragonlets but thorax all dark, with no pruinosity, and with no white bordering hindwing spots (which are shaped like coffee beans). Females and immatures of other small lowland drag-onlets (Red-faced, Red-mantled, Andagoya, and Chalk-marked) have dark markings on sides of abdomen interrupted and/or triangular in middle segments; only Coffee Bean has continuous dark and light stripes. **Habitat and Behavior:** Presumably breeds in forest ponds, although habitat not known in CR. Immatures at times very common along forest trails in South America, much more often seen than mature pruinose individuals, and from low in shrubs to well up in trees. Little known about dragonflies that spend much time in forest canopy, but this might be one. **Range:** Southern Pacific slope, to 100 m; so far known only from three Osa Peninsula specimens. Costa Rica to Argentina.

male

female

TL 34 mm; HW 25 mm. **Identification:** Has purplish face, dark brown eyes and thorax, white stripe on front of thorax and orange-brown, virtually unmarked, abdomen, colored unlike any other dragonlet though superficially like other species with hindwing spots such as Dwarf Glider (p. 348), with very different habits and habitat. In both sexes, hindwing conspicuously broader and rounder than in other dragonlets. **Habitat and Behavior:** Breeds well above ground in tank bromeliads in lowland forest. Males must seek out females high in trees, perhaps where they associate with bromeliads, rather than defending a territory at water as the great majority of skimmers do. This may be why sexes look alike. **Range:** Caribbean slope, to 100 m; so far known only from La Selva Biological Station. **Costa Rican endemic**.

immature male

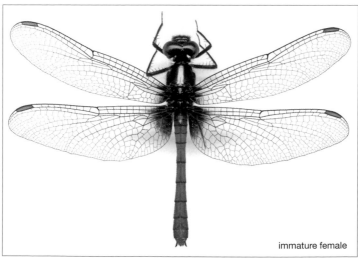

immature female

TL 32 mm; HW 23 mm. **Identification:** No other species breeds in its brackish-water environment, but as dragonflies wander, it must be compared with others. Males unique, although equally dark Metallic Pennant (p. 266) might overlap in distribution. Pennant has browner thorax, with some metallic markings, and more slender abdomen. Female/immature Seaside with orange-and-black striped thorax and orange-spotted abdomen, becoming darker with age and showing variation during process. Combinations occur such as black thorax with spotted abdomen or striped thorax with black abdomen. Spots vary in size, so that abdomen may be almost entirely orange. **Habitat and Behavior:** Breeding habitat in CR remains to be discovered, as usual habitats of coastal mangroves are limited and salt marshes absent from Caribbean coast. One of the few regional dragonflies with tandem oviposition. **Range:** Known from Caribbean coast of Central America, both north (Belize) and south (Bocas del Toro, Panama) of Costa Rica, but no records from country at present; also unknown in Honduras and Nicaragua. Coastal Nova Scotia to Venezuela, West Indies, and west coast of Mexico. Also inland in southwestern US. (Tropical form, which should occur in Costa Rica, is subspecies *Erythrodiplax berenice naeva*, slightly smaller than temperate zone *E. b. berenice* with limited wing markings and greater likelihood of mature female becoming pruinose gray.)

male

female

TL 42 mm; HW 29 mm. **Identification:** Mature males unique, all dark with black bands across each wing. Immatures of both sexes yellowish to pale tan, with dark brown abdominal markings, brownish stripes on either side of front of thorax, and vivid pale stripe from between wings to prothorax. Females and immatures distinguished from most other dragonlets by size and from very similar Black-winged Dragonlet (p. 297) by lateral abdominal markings pointed in front. Abdominal markings become duller in mature females, and some become extensively pruinose. Minority of females andromorphic, developing dark wing bands and dark body coloration, looking much like males. Wing bands in males and andromorph females at first pale, then darkening to black. Odd variants include wing bands smaller than normal in both sexes, even reduced to spots less than half area of normal bands. Many females at Los Chiles had such wing spots. **Habitat and Behavior:** Temporary and permanent ponds or open areas in slow streams; always with much emergent grasses and sedges, even densely vegetated. Much more common on Caribbean side than Black-winged. Although both occur countrywide, rarely if ever found in numbers together. At one marsh, Band-winged fiercely territorial to the few Black-winged present. Like Black-winged, Band-winged can be abundant in forest in dry season in immature state. **Range:** Both slopes, to 800 m. Southern US to Argentina, West Indies.

male

female

andromorph female

TL 40 mm; HW 30 mm. **Identification:** Mature males unique, with mostly black wings. In younger individuals, forewing black in middle with clear base, hindwing black in basal two-thirds. Band-winged Dragonlet (p. 295) superficially similar, but both wings black from nodus to stigma. Females and immatures very much like Band-winged, but markings on side of abdomen squared off, lacking pointed front that characterizes Band-winged. Minority of females andromorphic, with wings colored just like those of males. Immatures (predominant in dry season) straw-colored, camouflaged like dead grass, wings clear with dusky tips. Small percentage of heteromorph females have dark spot at base of hindwing (not present in Band-winged) and/or all wings with dusky tips. Some old females become extensively pruinose. Black-winged and Band-winged also distinguished by wing venation, Black-winged with two cell rows instead of one in medial planate (see illustration p. 11). These two large species the only dragonlets with two rows in radial planate. **Habitat and Behavior:** Ponds and marshy streams in open. One of the most abundant dragonflies in CR, especially on Pacific slope. Also locally common in Central Valley and well down onto Caribbean side in Cartago Province. Band-winged uncommon in both areas. Much less common and local at lower elevation on Caribbean side, absent from some areas where only Band-winged occurs. Adults spend dry season as immatures in woodland, and when rains begin in early summer quickly take on mature coloration and leave woodland to arrive at ponds everywhere. During Guanacaste study in 1967, sky became full of Black-winged Dragonlets moving over countryside well above ground just as rains came. At that time, every flooded puddle or series of cow footprints had numerous black-winged males hovering over it, with territories of 1 to 2 square meters. Many seen scattered along fences during midday heat, obelisking with abdomen straight up. **Range:** Both slopes and Central Valley, to 1400 m; uncommon and local on Caribbean slope. Western Mexico to Ecuador.

male

female

andromorph female

female variant

MEADOWHAWKS   genus *Sympetrum*

Two species. These small skimmers are typical of higher elevation, which is appropriate given their temperate-zone origin. Males of the Costa Rican species are mostly red, with two pale markings on each side of the thorax and dark markings at the wing bases. Pairs oviposit in tandem.

Cardinal Meadowhawk   *Sympetrum illotum*

TL 39 mm; HW 27 mm. **Identification:** Small red dragonfly of higher elevations with reddish wing veins, a pair of cream spots low on each side of thorax, and tiny dark spot at hindwing base. Females brown (abdomen red in some), and wing veins less reddish than in males. Red dragonflies such as tropical king skimmers, pondhawks, and redskimmers all larger as well as uncommon in or absent from its altitude range. Red-tailed Pennant (p. 264) somewhat similar but without reddish in wings or cream lateral thoracic spots. Those spots not present in any other skimmers but Talamanca Meadowhawk (p. 300). **Habitat and Behavior:** Temporary and permanent ponds in open areas. Oviposits in tandem but females also oviposit alone, hovering in one spot and methodically tapping water, rising slightly between taps; may also drop eggs from above water or drag abdomen through floating vegetation. **Range:** Both slopes, 1200–2600 m. Southwestern Canada and highlands of southwestern US to Panama.

male

female

299

TL 41 mm; HW 30 mm. **Identification:** Distinguished from Cardinal Meadowhawk (p. 299) by brown instead of red head, red on front of thorax, and dark wing veins. Black tibia and distal end of femur also distinctive. Female abdomen reddish to orange, brighter than many female Cardinal. Talamanca and Cardinal are the only two red dragonflies expected at such high elevations. **Habitat and Behavior:** Small ponds and pools in small streams in open areas. Perch on twigs or ground and hover over water. Pair oviposited by facing grass tufts from less than 30 cm away and tapping water every few seconds. **Range:** Central and Talamanca highlands, 1900–2700 m. **Costa Rican endemic**.

male

female

TROPICAL DASHERS   genus *Micrathyria*

Thirteen species. These small to medium skimmers are common at still waters everywhere. This is the most diverse skimmer genus in Costa Rica, and as many as four species can occur together at some sites. Along with dryads (*Nephepeltia*), they are the only Costa Rican skimmers with two bridge crossveins (see wing diagram, p. 11), easily seen in the hand or in photos. The following characters are found in all species. In immatures, eyes are red above and gray-tan below; in mature males, they turn brilliant green or blue-green; the same transition perhaps takes place in all females of advanced age. All have a white face below a dark metallic frons. Males and most females have a black abdomen with prominent, paired pale spots on S7, some with additional markings on middle abdominal segments. All have a pair of pale frontal stripes on the front of the thorax and horizontal stripes above those and on the antealar carina; in front view, the horizontal stripes can resemble a pair of tiny wings. The thorax of all species is pruinose between the wing bases in males, and pruinosity may extend onto the abdomen base. Thorax of females and immatures of both sexes is marked with a pattern of complex stripes; males in some species become increasingly dark, some with thorax and abdomen base becoming pruinose with maturity. Females usually have more extensive pale markings on the abdomen, and the thorax and even the abdomen may become pruinose, looking more like that of males as they age.

Individuals perch out in the open and often droop their wings forward when relaxed. Males are quite aggressive with their own and other similar species. Although almost all skimmers oviposit by dropping their eggs into water, in some tropical dashers the mode is *epiphytic*, the female hovering close or landing to deposit eggs on stems and leaves. The most similar skimmers are setwings (*Dythemis*) and dryads (*Nephepeltia*). Setwings are large and slender, and mature males have blue or reddish eyes, with no pruinosity between wings. Dryads are small, the front of thorax either with pale spots or entirely dark.

Black Dasher   *Micrathyria atra*

TL 41 mm; HW 31 mm. **Identification:** Only *Micrathyria* with narrow, entirely black (mature) or mostly orange (immature) abdomen, no hint of club; rather large. Fairly long, narrow (because of narrow abdomen) spots on S7 of males distinctive, additional spots on S8 in some individuals even more so; also, spots more likely to be orange than in any other dasher. Male Morton's Skimmer (p. 255) very similar to orange Black Dasher, with striped thorax and mostly orange abdomen. Orange on abdomen tapers to point in middle of S7 in Morton's, while Black has pair of broader, isolated markings on S7. Note female Dusky Dasher (p. 308) with dull orange unpatterned abdomen; pattern much more obscure, unlike vivid black and orange of immature Black. **Habitat and Behavior:** Ponds and marshes, usually associated with forest and often under canopy. Active only during sunny periods but may become common at water even in early morning. Males hover 1–3 m above water, hovering much more frequently than other tropical dashers. Oviposition by either tapping water, as is typical of skimmers, or by hovering a meter above it and dropping small bunches of eggs, often turning 180° after each bunch. **Range:** Both slopes, to 1200 m. Guatemala to Argentina.

male

female

orange male

TL 36 mm; HW 27 mm. **Identification:** The most common large dasher; lateral thoracic stripes parallel, uninterrupted, and not joined with other markings, thus showing less complex pattern than that in most other dashers. Parallel stripes and large posterior hamules of males shared only with three other species. Of these, Even-striped Dasher (p. 304) very similar but with larger hamules extending farther forward and tilted upward. Females even more similar, might be distinguished by slightly smaller spots on S7 in Even-striped. Other two with large hamules are Swamp (p. 305) and Forest Pond (p. 306) Dashers, differing from Three-striped by wider posterior dark lateral thoracic stripes; the stripes are partially fused in Swamp, resembling a single stripe. Female Three-striped rarely have dusky wing tips. **Habitat and Behavior:** Forested ponds and swamps, usually in sun. Males at water tend to perch higher than other dashers. **Range:** Pacific slope, to 300 m; also a few records from Caribbean lowlands. Texas and Florida to Ecuador and French Guiana, West Indies.

male

female

TL 35 mm, HW 25 mm. **Identification:** Very similar to Three-striped Dasher (p. 303), especially evenly striped thorax; males distinguished in side view by slightly shorter, straighter cerci and distinctly longer posterior hamules that extend well forward and curve upward. S7 spots average slightly smaller, more likely to be rounded at ends. Often has noticeable contrast between greenish or bluish thorax and yellowish to orange mid-abdominal markings, especially notable in females; spots on S7 usually greener or paler than other abdominal markings. Even-striped has forewing triangle one-celled, Three-striped two-celled; evident in hand and photos. **Habitat and Behavior:** Ponds in forest. Females in light gaps along forest trails. **Range:** Southern Pacific and Caribbean slopes, to 900 m. Southern Mexico to Peru and Venezuela.

male

female

TL 35 mm; HW 28 mm. **Identification:** Pattern on thorax unique: note width of pale side stripe and also dark stripe behind it that looks wider than it is due to near fusion of two stripes. Prominent male hamules distinct from hamules in Three-striped (p. 303) and Even-striped Dashers in being blocky; posterior hamules do not extend forward very far, while other two species have longer forward extension visible from side. Three-striped and Even-striped also have dark and pale thoracic stripes about even in width. Also much like Forest Pond Dasher (p. 306), which occurs above 700 meters, with perhaps no overlap in elevation. Females only CR dasher with prominent dark wing tips, although some female Three-striped and Even-striped show hint of this. **Habitat and Behavior:** Swamps and ponds within forest. Like Three-striped, males perch at head height and higher on tree branches in sun or shade. **Range:** Caribbean lowlands, to 100 m. Guatemala to Peru and Brazil.

male

female

TL 38 mm; HW 31 mm. **Identification:** Most similar to Swamp Dasher (p. 305) in size and color pattern but probably no overlap in elevation. Males distinguished by slightly wider pale stripes on front of thorax in this species, as well as longer posterior hamules. Females distinguished from Swamp by lacking dark wing tips. Also quite similar to Three-striped Dasher (p. 303) and overlaps with it. Thoracic-stripe pattern should distinguish both sexes. **Habitat and Behavior:** Marshy ponds in forested landscape. **Range:** Highlands, 700–1450 m. Costa Rica to Venezuela and Ecuador, also Argentina.

male

female

TL 31 mm; HW 23 mm. **Identification:** Males of three *Micrathyria* have a pruinose gray thorax; of them, Spot-tailed and Dusky (p. 308) Dashers usually have single pair of abdominal spots on S7, Peten Dasher (p. 314) with multiple markings. In Spot-tailed, spots end in points; rounded in Dusky. Thornbush Dasher (p. 313) also similar in size, with all segments marked, spot on S7 rounded or square. Square-spotted Dasher (p. 309) with squarish spots on S7 much larger than markings on preceding segments. S7 of Peten and Little Swamp (p. 315) Dashers also marked with long triangles as in Spot-tailed, but both smaller, and darker abdomen has less extensive pale markings at base than in Spot-tailed. Females distinguished from most others by conspicuous abdominal pattern of backward-pointing triangles that are wider on S7 than on S6. Oldest females develop pruinose abdomen and green eyes. In other females with somewhat similar markings, marking on S7 is not a triangle. **Habitat and Behavior:** Ponds and marshes. Males perch low when at water, higher on branches when away. Females may perch well up in trees, usually at forest edge. Oviposition epiphytic, laying eggs on underside of floating leaf or stem while perched on it. **Range:** Both slopes and Central Valley, to 1200 m. Texas and Florida to French Guiana and Ecuador, West Indies.

male

female

TL 32 mm; HW 24 mm. **Identification:** Males with pruinose thorax colored much like similarly sized Spot-tailed Dasher (p. 307) and obviously smaller Peten Dasher (p. 314) but a bit darker gray; some males even look shiny black on thorax. Lacks extra abdominal spots of Peten. Usually with spots on S7 rounded at ends rather than pointed. Male Dusky often shows a tiny touch of orange at hindwing bases, absent in male Spot-tailed. Female more distinctive than male; only dasher with plain light reddish-brown abdomen, not as dramatically orange and black as in female Black Dasher (p. 301). Dusky also may have brownish wing tips, but much less obvious than dark, sharply demarcated tips of female Swamp Dasher (p. 305). Female Dusky also often with faint orange wash on basal third of wings, unique among *Micrathyria*. Both female and immature Dusky show two short streaks extending down from top of thorax on either side of midline; no trace of them in Spot-tailed. Not to be confused with the paired isolated markings in same area on Fork-tipped Dasher (p. 311). **Habitat and Behavior:** Ponds in forest or open areas. Like Spot-tailed, oviposits after landing on surface vegetation. **Range:** Pacific slope, to 300 m. Western Mexico to Venezuela.

male

female

TL 32 mm; HW 24 mm. **Identification:** Males have best-developed club of any regional dasher, very prominent from above, with very large square spots on S7. Females distinguished by broad yellow-orange stripes back to S6 and squareish spots on S7 much larger and paler than other species. See very similar Dark-fronted Dasher (p. 310). Also compare female Thornbush (p. 313) and Fork-tipped (p. 311) Dashers. **Habitat and Behavior:** Ponds, marshes, and roadside ditches in sun; more common in vegetation than over open water. One of the most common and widespread dashers. Oviposition by dropping eggs from just above surface or crawling on floating vegetation and dragging abdomen while extruding eggs in a linear mass. **Range:** Both slopes, to 1200 m. Central Mexico to Argentina.

male

female

TL 36 mm; HW 26 mm. **Identification:** Males similar to males of much more common Square-spotted Dasher (p. 309), averaging slightly larger and with spots on S7 slightly smaller, often less square. In side view note male epiproct extends just to prominent tooth on underside of cercus, extending considerably farther beyond that point in Square-spotted. In both sexes, sides of thorax look somewhat pale in contrast with dark front. Long abdomen sometimes curved downward toward tip, unique in the genus. Also called Bow-tailed Dasher. **Habitat and Behavior:** Large ponds in forested habitat. Seems to be quite local and scarce in CR. Behavior much like Square-spotted but oviposition by tapping water surface. **Range:** Caribbean slope, to 900 m; more common at higher elevation. Costa Rica to Argentina.

male

female

TL 32 mm; HW 23 mm. **Identification:** Males identified at close range by long, pointed diverging cerci that give the species its common name; there are apparently two color morphs, see below. Striped males readily distinguished from all other *Micrathyria* by extra pair of markings on front of thorax, short streaks on either side of midline. Other species with slender, slightly clubbed abdomen have spots on basal segments, while Fork-tipped has streaks. Somewhat similar species such as Even-striped (p. 304) and Swamp (p. 305) Dashers have different thoracic patterns. Spot-tailed (p. 307) and Dusky (p. 308) Dashers lack markings on mid-abdomen, and Spot-tailed develops pruinose thorax. Female Fork-tipped distinguished from rather similar Square-spotted Dasher (p. 309) by smaller markings on S7 and no markings on S6; oddly, another male morph, so far found only at Los Chiles and La Selva, has unstriped glossy thorax with scattered pale markings looking much like a dryad, *Nephepeltia*. Dryads are smaller, and Spine-bellied Dryad has pale spots on front of thorax. Check the appendages! **Habitat and Behavior:** Sunny ponds with emergent vegetation, bordered by forest or not. Males defend territories usually from low perches. Female oviposits by thrusting abdomen under leaf of floating plant and withdrawing it slowly, leaving double row of eggs plastered beneath it. **Range:** Both slopes, to 100 m. Guatemala to Peru and Brazil.

male

female

male glossy morph

TL 34 mm; HW 27 mm. **Identification:** Large, flask-shaped (narrow above, bulging below) pale stripe on each side of thorax underneath forewing characteristic, as is male abdomen, most heavily spotted of any dasher. Female abdomen similar to that of Spot-tailed Dasher (p. 307), with prominent markings from S1–6, but markings on S7 broad, not triangular. **Habitat and Behavior:** Open vegetated ponds; both sexes often in nearby woodland. Oviposition by dropping eggs from above water or laying them while crawling across slightly submerged vegetation. **Range:** Pacific slope, to 700 m; most common in Guanacaste region. South central US to Panama, Greater Antilles.

male

female

TL 25 mm; HW 20 mm. **Identification:** Quite small, males look like smaller version of male Spot-tailed (p. 307) or Dusky (p. 308) Dashers but with spots on middle abdominal segments as well as S7. Look for pale tips of male cerci, only on this and, even less visible, Little Swamp and Dusky Dashers. Also note that some individuals lack extra spots and pale cercus tips, so checking identity (in hand if possible) of all small *Micrathyria* in an area is more likely to yield correct identifications. Females distinguished by small size and pale triangles of equal size from abdomen base to S7; female Little Swamp almost identical but markings on middle segments average narrower. Female dryads (*Nephepeltia*), also tiny and superficially similar, have front of thorax unmarked or with pale spots rather than stripes as in *Micrathyria*. **Habitat and Behavior:** Common at edges of small to large ponds, in forest and open areas, where males perch on tips of aquatic vegetation. Epiphytic oviposition observed. **Range:** Northern Caribbean slope, to 100 m; known only from Boca Tapada. Eastern Mexico to Costa Rica.

male

female

TL 24 mm; HW 20 mm. **Identification:** Tiny dasher, males with brightly marked metallic thorax and no markings on abdomen except S7 spots. Mature male Peten Dasher quite similar but with pruinose thorax, pointed spots on S7, and additional markings on S5-6. Pale tips on cerci more obvious in Peten. Male Glossy-fronted and Spine-bellied Dryads lack frontal stripes on thorax (may have spots) and have longer, entirely black cerci. **Habitat and Behavior:** Ponds in forest, where males perch low over water in sun. **Range:** Both slopes, to 300 m. Guatemala to Argentina.

male

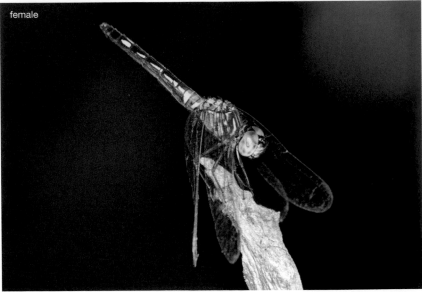

female

genus *Micrathyria* (p. 301)

selected male appendages in lateral view

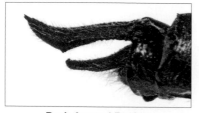

Dark-fronted Dasher
Micrathyria catenata (p. 310)

Square-spotted Dasher
Micrathyria ocellata (p. 309)

selected male hamules from right side

Even-striped Dasher
Micrathyria dictynna (p. 304)

Three-striped Dasher
Micrathyria didyma (p. 303)

Swamp Dasher
Micrathyria laevigata (p. 305)

Forest Pond Dasher
Micrathyria venezuelae (p. 306)

DRYADS   genus *Nephepeltia*

Two species. With bright green eyes, white face, and slender black abdomen with big spots on S7, these small dragonflies resemble tropical dashers (*Micrathyria*), and even perch like them, with wings depressed. They can be distinguished in hand or in photos of wings by the last antenodal before the nodus being complete between the costal and subcostal veins and again between the subcosta and radius, while in tropical dashers that vein extends only between the first two, with an empty space behind (called "antenodals incomplete" in keys). The front of the thorax in dryads is glossy green-black, either unmarked or with a pair of pale spots; tropical dashers all show stripes there when they are not covered by pruinosity. See Fork-tipped Dasher, a larger species, for an uncommon color morph with glossy thorax. Dryad male cerci are long and curve upward toward the tips, unlike any tropical dasher.

Glossy-fronted Dryad   *Nephepeltia flavifrons*

TL 25 mm; HW 18 mm. **Identification:** Entirely black metallic front of thorax distinguishes both sexes from all other similar dragonflies except those Spine-bellied Dryads (p. 318) that lack spots on front of thorax and some Fork-tipped Dashers (p. 311). Both sexes of Spine-bellied become pruinose on sides of thorax with maturity, not so in Glossy-fronted. Females of small species of tropical dashers have pale stripes on front of thorax. **Habitat and Behavior:** Ponds, usually associated with forested areas. Males perch low over water. **Range:** Caribbean slope, to 100 m. Eastern Mexico to Argentina.

male

female

TL 24 mm; HW 18 mm. **Identification:** No other small dragonfly with paired pale spots on front of thorax, though some males lack spots. Males without spots resemble Glossy-fronted Dryad (p. 317) except for extensive pruinosity on sides of thorax. Unique projecting spine on underside of male thorax difficult to see. Females of these two species very similar but Spine-bellied thorax usually shows spots on front and pruinose sides with maturity, as well as more heavily spotted abdomen. Small species of tropical dashers very similar but all but glossy morph of Fork-tipped (p. 311) have stripes on front of thorax. **Habitat and Behavior:** Ponds and marshes in or out of forest. Males perch low over water, also up into shrubs to 2 m. **Range:** Both slopes, to 600 m. Guatemala to Argentina.

male

female

LEAFSITTERS genus *Oligoclada*

Two species. Leafsitters are distinctive among small skimmers in their greenish eyes, long legs, leaf-perching behavior, and overall gray to black coloration (abdomen weakly metallic). Metallic Pennant, with brown eyes, perches on plant tips. Dark dragonlets such as Andagoya have more and bluer pruinosity, dark brown eyes, and pale terminal appendages. The relatively long legs of leafsitters are perhaps to elevate them above the leaves on which they usually perch; apparently, they are not adapted for perching on twigs or stems. Mature males are black but with pale pruinose blue-gray on the sides of the thorax, between the wings, and on the abdomen base, and in contrast to tropical dashers and dryads, the abdomen lacks pale markings. Females are rarely seen and have not been described for either CR species. Capturing and photographing a mated pair would be a significant contribution to our understanding of these species. Immatures of both sexes are probably like other leafsitters, with a reddish thorax and black abdomen with a reddish base.

Sunshine Leafsitter *Oligoclada heliophila*

TL 27 mm; HW 20 mm. **Identification:** Males virtually identical to Shadowy Leafsitter (p. 320) but have pair of tiny teeth on rear margin of occiput and larger teeth on lower side of cerci. Also, Sunshine cercus arched near middle, not so in Shadowy. In close side view, male hamules might be visible. In this species, note small backward-directed hook on rounded hamule, while Shadowy has wider hamule with small forward-directed hook at rear. Females not described for either species, probably colored like male but might show reddish basal abdominal segments. Forewing triangle crossed in male Sunshine, not crossed in Shadowy; possibly useful in distinguishing females. **Habitat and Behavior:** Small streams in sunny pastures and clearings. **Range:** Caribbean and southern Pacific slopes, to 200 m. Costa Rica to Peru and Venezuela.

male

TL 25 mm; HW 21 mm. **Identification:** See Sunshine Leafsitter (p. 319). **Habitat and Behavior:** Small streams and ponds in forested areas, light gaps on forest trails. Said to differ from Sunshine Leafsitter by usually being in shade. **Range:** Caribbean slope, to 300 m. Guatemala to Ecuador and Venezuela.

male

TL 25 mm; HW 21 mm. **Identification:** Nothing else like this small dark species with dark wings, brilliant reflective blue on upper surface. Much too small to be confused with morpho butterflies in flight, but similar to several species of *Heliconius* butterflies (even speculation that they mimic these distasteful butterflies, although they look more like a skipper when perched). Only dragonfly in CR that commonly perches with wings closed; butterflies have much larger wings relative to abdomen length. Female wings less strikingly blue above, with more conspicuous white bands across them. Also called Rainforest Bluewing. **Habitat and Behavior:** Breeds in forest-bordered ponds of all sizes. After flight, almost always lands with wings closed over thorax, unique among New World dragonflies and presumably to hide blue color from potential predators or rival males. Males after short while depress wings well downward, leaving blue upper surface conspicuous. Frequent chasing and circling interactions with other males. Females oviposit among grasses, with or without guarding male. Both sexes perch high in forest trees with wings depressed, looking like little parasols. May even perch with forewings open, hindwings closed. Females also perch lower in clearings along forest trails. **Range:** Lowlands of Caribbean slope, to 700 m. Costa Rica to Bolivia and Brazil.

male

female

SETWINGS genus *Dythemis*

Two species. Setwings are skimmers of streams and swamps with a finely striped thorax; they perch on twigs with wings very often considerably depressed ("set"). Costa Rican species have a long, slender, streaked or spotted abdomen and just a touch of color at the hindwing bases. The underside of S9 in females shows a projecting curved ridge, prominent in side view, perhaps to divide extruding egg masses.

Blue-eyed Setwing *Dythemis nigra*

TL 37 mm; HW 29 mm. **Identification:** Blue eyes of mature individuals and long, slender abdomen with most prominent spot on S7 distinctive. No dasher (*Micrathyria*, pp. 301–316) has such a slender abdomen or such blue eyes, and most similar dashers (Three-striped, p. 303, and Even-striped, p. 304) have a less complexly striped thorax. Younger individuals with reddish eyes could be confused with Brown Setwing but note metallic frons and spotted rather than streaked abdomen. Some females develop prominent dark wing tips. **Habitat and Behavior:** Ponds and streams. Males defend territories over stream pools and even at rain puddles on forest trails. Females perch high on leaves and twigs at clearing edges. Surprising variation in oviposition: one female oviposited erratically over open water, flying low and tapping at intervals, while another oviposited slowly and methodically at edge of grasses. **Range:** Both slopes, to 1000 m. Northern Mexico to Argentina.

male

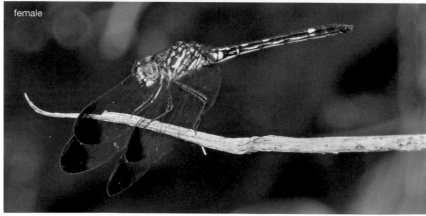
female

TL 38 mm; HW 28 mm. **Identification:** Pattern much like Blue-eyed Setwing but eyes brick red above and body reddish and brown where that species is blue and black. Abdominal markings long triangles or streaks rather than spots. Dark wing tips when present in female less conspicuous and extensive than in female Blue-eyed. Superficially like dashers but distinguished by brown eyes (note, however, that immatures of perhaps all blue- and green-eyed species have red-brown eyes). **Habitat and Behavior:** Streams and rivers, less commonly ponds and lakes. Territorial males perch up to 2 m above water. **Range:** Both slopes, to 1000 m. Northern Mexico to Peru and Venezuela, Lesser Antilles.

male

female

STREAMSKIMMERS genus *Elasmothemis*

Two species. These dragonflies are stream dwellers. Males cruise up and down streams like clubskimmers and sylphs but perch like setwings, on leaves and twigs over the water with wings depressed. Uniformly reddish or orange coloration, some individuals with evenly spaced faint dark markings on the abdomen that distinguish them from any other stream species.

Ruddy Streamskimmer *Elasmothemis cannacrioides*

TL 46 mm; HW 35 mm. Identification: Slightly smaller and more reddish than Apricot Streamskimmer but sometimes more yellowish than reddish. Cerci blackish (red in some other reddish species). Abdomen not bright red as in many superficially similar species, and the other red species are not usually seen flying back and forth over running water as this species does. Also called Golden Streamskimmer. **Habitat and Behavior:** Males fly beats over forest streams and rivers with moderate currents; especially favors gravel riffles. Also hovers for long periods over water, then flies to new spot and hovers again. Females, accompanied by males, seen dragging abdomen tip along rootlets in stream to oviposit. Laying eggs directly on plants is known as epiphytic oviposition, unusual in dragonflies. Eggs shed in stringlike filaments, also unusual for this family. **Range:** Both slopes, to 500 m. Eastern Mexico to Peru and Argentina.

male

female

TL 48 mm; HW 38 mm. **Identification:** Very similar to Ruddy Streamskimmer, but a bit larger and with face brighter yellow. Overall coloration varies from yellow-orange to orange, with contrasting black cerci. Usually with less black on abdomen than Ruddy. Anterior lamina of male genitalia pale in this species, dark in Ruddy; might be visible at close range. **Habitat and Behavior:** Presumably much like that of closely related Ruddy Streamskimmer. **Range:** Northern Caribbean slope, to 400 m. Eastern Mexico to Costa Rica.

male

CLUBSKIMMERS   genus *Brechmorhoga*

Five species. These medium to large skimmers have pale blue to greenish eyes, a dark face that is dark metallic blue above and white below, dark brown thorax with a pair of pale stripes on the front, and two pale yellowish, cream, or pale green stripes on the sides, with usually a narrow third stripe between them. Males have a long, slender abdomen with an expanded clublike tip. S7 has a pair of pale spots, like those in many other stream dragonflies, and the shape of the spot is unique to each species. Females are similar but with a duller face and pale markings on more segments of a less expanded abdomen; the forewing tips are sometimes suffused with amber, and wings become entirely brownish in oldest individuals.

Males fly beats up and down streams, chasing one another during encounters. Between flights, they perch flat on rocks or on leaves with their abdomen lower than their thorax. Both sexes fly over clearings to forage on small insects, sometimes in small groups mixed with sylphs and others. Away from water they hang up more or less vertically, like darners. Except in the hand, with a close view of the male hamules, they can be difficult to identify—even more difficult in the case of females. Fortunately, usually no more than two species occur on a given stream. In the hand, note that males of the larger species (Masked and Amber-banded, both at middle elevations) have three cell rows between the anal loop and the hind margin of the hindwing, while smaller ones (Little, Slender, Vivacious, all more common in lowlands) have two rows; most females have three rows.

Little Clubskimmer   *Brechmorhoga nubecula*

TL 43 mm; HW 30 mm. **Identification:** Smallest clubskimmer, similar in size to sylphs but with abdomen neither widely expanded nor very slender throughout as in some sylphs. Middle lateral thoracic stripe usually more obvious than in other clubskimmers, which at a glance appear to have two stripes. Abdomen relatively shorter than in Slender and Vivacious Clubskimmers, markings on S7 less prominent. In hand note that males have only a single tooth on long, curved cerci. Females with unusual broad, flattened cerci rounded at tips; slender and pointed in all other female clubskimmers. Also called Bachelorette Clubskimmer. **Habitat and Behavior:** Small streams and rivers in forest. Much less frequently seen than other CR clubskimmers, females usually in foraging flight away from water. One female seen ovipositing in shallow water of swift shaded stream, another in a roadside ditch with swift flow. Where are the males? **Range:** Both slopes, to 700 m. Belize to Argentina.

male

female

TL 46 mm; HW 36 mm. **Identification:** Abdominal markings yellower than in other clubskimmers; marking on S7 bold and bright yellow-orange all around, taking up entire segment, and abdomen shorter and stouter than in others. Only female Masked Clubskimmer (p. 328) shows as much yellow, but abdomen slightly longer and narrower and underside of S7 dark in that species. **Habitat and Behavior:** Rocky streams and rivers in forested landscapes and pastures. While cruising on streams, males often land on low vegetation, rocks, and bare ground; other clubskimmers usually land on vegetation, typically higher up. Males closely guard females as they lay eggs over shallow riffles. One female oviposited by hovering and slowly changing position, tapping water every 5–10 seconds. **Range:** Both slopes, 700–1700 m. Highlands of southern Mexico to Peru and Venezuela.

male

female

TL 52 mm; HW 37 mm. **Identification:** Large; males with abdomen relatively shorter and broader at S7 than in Slender and Vivacious Clubskimmers. Females, with almost entire S7 yellow, difficult to distinguish from female Amber-banded (p. 327), with which they often occur, but Masked has abdomen relatively more slender, and yellow on S7 more obviously bordered by black at rear of segment. **Habitat and Behavior:** Males fly up and down over small to medium streams in forested and semi-open habitats, occasionally hovering. Females oviposit methodically in swift water and open gravel pans after brief aerial copulation, guarded by males hovering above them. **Range:** Both slopes, 600–1600 m. Arizona to Bolivia.

male

female

TL 46 mm; HW 30 mm. **Identification:** One of two smaller clubskimmers with relatively long abdomen and two cell rows between anal loop and hindwing margin. Note extension of white face color up along eyes in this species, not in Vivacious (p. 330). Male hamules straight in middle, then curved toward tip; strongly curved throughout in Vivacious. Paired pale dorsal markings on S3 extend length of segment in Slender but much smaller or absent on front half of segment in Vivacious. Females indistinguishable in field, in hand note abdomen longer than hindwing in this species, about same length or shorter in Vivacious. Look for longitudinal green stripe on upper side of S2 joining vertical basal stripe in Slender; these stripes separate in Vivacious. Subgenital plate of this species with deep U-shaped notch, only shallowly notched in Vivacious. Forewings suffused with amber tips in some females, entire wings slightly brownish with age. **Habitat and Behavior:** Streams and rivers, usually forested. **Range:** Pacific slope, to 1100 m; Caribbean slope, to 600 m. Arizona and Texas to Peru and Brazil.

male

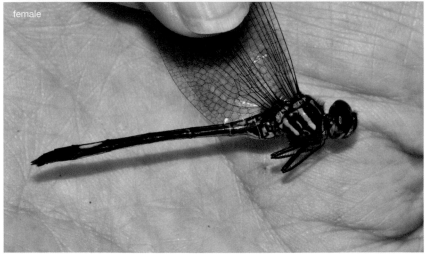

female

TL 49 mm; HW 31 mm. **Identification:** One of two clubskimmers (see Slender, p. 329) with relatively long, slender abdomen with narrow spots on S7 and two cell rows between anal loop and wing margin. Rare individuals have brownish tips to forewings, but entire wings suffused brownish with age. A bit darker brown than Slender Clubskimmer; see that species for other differences. **Habitat and Behavior:** Streams and rivers. **Range:** Both slopes, to 1400 m. Northern Mexico to Argentina.

Little Clubskimmer
Brechmorhoga nubecula (p. 326)

Masked Clubskimmer
Brechmorhoga pertinax (p. 328)

Slender Clubskimmer
Brechmorhoga praecox (p. 329)

Amber-banded Clubskimmer
Brechmorhoga rapax (p. 327)

Vivacious Clubskimmer
Brechmorhoga vivax

SYLPHS   genus *Macrothemis*

Eleven species. Males of these slender dragonflies fly up and down streams, sometimes landing flat on leaves or rocks. When away from the water, they usually perch by hanging up. A striped thorax and patterned abdomen, usually with expanded markings on S7, is typical of most species. The pale thoracic stripes of some species are so wide the resulting pattern looks like dark stripes on a pale background. Males of most have clubbed abdomens, not evident in females; three (Attenuate, Straw-colored, and Delicate) have abdomens slender throughout. Females of some species (Delta-fronted, Ivory-striped, Jade-striped, and White-tailed) have characteristic orange to brown markings on their wings, though these are not always present. As in setwings, S9 in female is curved and ridged below like a sled runner, often prominent from the side. All are smaller than the somewhat similar and closely related clubskimmers, except the two largest species (Attenuated and Straw-colored), both of which are slender. Both sexes indulge in swarm-feeding in forest clearings and over roads at any height, sometimes along with clubskimmers and others. Past feeding swarms at La Selva included as many as five species: Delia, Delta-fronted, Ivory-striped, Jade-striped, and Delicate.

Gold-spotted Sylph   *Macrothemis aurimaculata*

TL 37 mm; HW 27 mm. **Identification:** Males with abdomen expanded into distinct club, with small pale markings at base and prominent pair of yellow spots on S7 rounded in front and straight at rear. Females may have small yellow spots on S1-6 and large spots like those of males on S7. Only sylph with yellow markings almost filling S7. Lateral markings of thorax distinctive, with middle of three stripes broken into spots. **Habitat and Behavior:** So far found only while feeding over open areas; presumably breeds in streams as do other sylphs and may be restricted to uplands. **Range:** Pacific slope in San Vito area, 1200 m; not seen since 1968. Guatemala and Costa Rica.

male specimen

female specimen

TL 36 mm; HW 27 mm. **Identification:** Distinguished by relatively short, narrow pale stripes on sides of thorax and narrow pale markings down much of abdomen, with small spots on S7. Anterior edge of forewings tinted brownish in some females. Only sylph with male cerci truncate at ends. **Habitat and Behavior:** Collected from foraging flocks over clearing in forest; presumably breeds or bred in nearby Rio Puerto Viejo. **Range:** Caribbean slope, to 100 m; common at La Selva Biological Station in 1967 but not seen since then. Southern Mexico to Ecuador and Suriname.

male

male

Delicate Sylph *Macrothemis musiva*

TL 37 mm; HW 23 mm. **Identification:** Slender dark abdomen with fine streaks, together with broad and irregular stripes on thorax, distinguishes both sexes of Delicate from all other sylphs except rather similar Attenuate Sylph. The latter is slightly larger and has triangular frontal stripes and more complete second lateral stripes on thorax; also, no pale markings on middle abdominal segments. Male cerci straight in Attenuate, curved down and then up toward tip in Delicate. **Habitat and Behavior:** Breeds in small forest streams, sometimes in feeding swarms in clearings. **Range:** Both slopes, to 1000 m. Eastern Mexico to Argentina.

male

female

TL 43 mm; HW 26 mm. **Identification:** Extremely slender abdomen distinctive in both sexes, especially middle segments; only Delicate Sylph similar. **Habitat and Behavior:** Breeding habitat forest streams and may prefer larger rivers. **Range:** Caribbean slope, to 100 m; apparently rare and so far found only at Los Chiles and La Selva Biological Station. Southern Mexico to Peru and Brazil.

male

female

TL 47 mm; HW 33 mm. **Identification:** Only sylph with narrow, clubless abdomen, except much darker and even more slender Attenuate Sylph and much smaller and darker Delicate Sylph. Attenuate also much less common and perhaps no overlap in range in CR. No other skimmer with similar abdomen, except perhaps Pin-tailed (p. 277) and Red (p. 278) Pondhawks, but pondhawks with expanded abdomen base and quite different behavior. Female with base of hindwings tinted brown and faint indication of similar color at tips of forewings. **Habitat and Behavior:** Breeds in larger rivers than those inhabited by other sylphs, commonly in open stretches. Males often perch flat on rocks or even on ground. Females oviposit by flying low and fast and tapping water every second or so. Both sexes in feeding swarms with other sylphs. **Range:** Pacific slope, to 400 m. Arizona and Texas to Argentina.

TL 37; HW 27. **Identification:** Males with combination of paired short triangles (delta spots) on front and four relatively large stripes and triangular markings on sides of thorax. Male cerci longer than in White-tailed (p. 338) or Deceptive (p. 340) Sylphs, both slightly larger. Spines on inside of male femur finer than in White-tailed. Female wings variable, but either totally shaded orange; with dark forewing tips; orange at base of both wings; or with both dark tip and orange base. Only other females with dark wing tips are Jade-striped, with all four wing tips dark. **Habitat and Behavior:** Breeds in small streams and flowing ditches, males in short back-and-forth patrol flights. Both sexes often seen in buoyant feeding flight over clearings up to treetop level. Dark forewing tips make for odd look in cruising females. **Range:** Both slopes, to 300 m. Northern Mexico to Argentina.

male

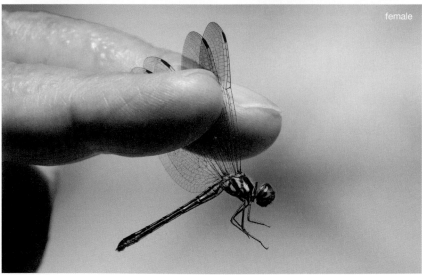

female

TL 39 mm; HW 29 mm. **Identification:** Characterized by four irregular well separated spots on each side of thorax and wide spots on S7 of expanded abdomen tip. Only sylph with large area of pruinosity from S7 to abdomen tip, but only a minority of males show this. Similar to Delta-fronted Sylph (p. 337), which is smaller and has markings on S7 not much wider than on preceding segments; spots on S7 much wider in White-tailed. Delta-fronted has markings on front of thorax shorter and those on sides of thorax longer. Male cerci longer and straight in Delta-fronted, shorter and slightly curved in White-tailed. Females have extreme wing bases brown, and both wings show faint orange wash and are darker on outer third of forewing. See also Deceptive Sylph (p. 340), very much like White-tailed but much less common. In both sexes of White-tailed, outer side of hind tibiae light brown; black in Deceptive and Delta-fronted. **Habitat and Behavior:** Breeds in streams and rivers, sometimes foraging far from water. More likely than other similar-looking sylphs to land on rocks. **Range:** Both slopes, to 1100 m. Arizona and Texas to Peru and Brazil.

male

female

pruinose male

TL 38 mm; HW 29 mm. **Identification:** Extremely similar to White-tailed Sylph (p. 338) in shape and structure. Markings on sides of thorax in both sexes larger, first and second spots overlapping by more than half their length, the second reaching the spiracle, while scarcely overlapping in White-tailed. Pale markings on S7 relatively narrow and twice as long as wide in Deceptive, not much longer than wide in White-tailed. Spines on male hind femur smaller than in White-tailed; hind tibiae black in Deceptive, light brown in White-tailed. Female wings with faint to distinct orange shading from base to nodus, clear at tip; female White-tailed may have some color beyond nodus. See also Delta-fronted (p. 337). **Habitat and Behavior:** Breeding habitat probably small forest streams. Rare in CR compared to White-tailed. **Range:** Both slopes, 400–800 m; so far found only at Pascua in Rio Reventazón watershed and Brujo in Rio Térraba watershed. Nicaragua to Ecuador.

TL 35 mm; HW 26 mm. **Identification:** Strongly clubbed males distinguished by single prominent white stripe on each side of thorax under forewing; short triangle on front a bit smaller than those on other sylphs, looking like headlights on individual in flight. Abdominal markings end in small spots on S7 rather than the conspicuously large spots of other clubbed species such as Delta-fronted, White-tailed, and Deceptive Sylphs, which also have lateral thoracic markings unlike the stripe in Ivory-striped. Male cerci more obviously curved upward toward tips than those of other sylphs, except Ski-tipped. Some females with light orange shading in wings, especially toward leading edges, and tiny dark markings at wing bases shared with female White-tailed. **Habitat and Behavior:** Breeds in streams, feeding aerially in nearby open areas. **Range:** Both slopes, to 400 m. Texas to Argentina.

male

female

341

TL 34 mm; HW 25 mm. **Identification:** Wide pale thoracic stripes and mostly black abdomen with rather small round spots on S7 distinguish this species. Same spots in Gold-spotted, Deceptive, and White-tailed Sylphs are much larger, and latter two have mostly dark thorax with pale markings. Jade-striped appears to be the only sylph in which some females have all wing tips conspicuously dark, distinctive in the air. Others have orange wing bases out to area of triangles, more rarely both colored wing bases and tips, Delta-fronted with dark forewing tips only. Note very similar Ski-tipped Sylph. **Habitat and Behavior:** Breeds in forested streams. Common, especially females, in small feeding swarms in sunny open areas. Both sexes typically perch flat on leaves. **Range:** Both slopes, to 1000 m. Texas to Ecuador and Venezuela.

male

female

TL 34 mm; HW 25 mm. **Identification:** Looks exactly like Jade-striped Sylph, but males have cerci much more strongly curved upward past ventral teeth, anterior lamina more prominent, projecting forward more, and more rounded (more angular in Jade-striped), and spines on inner surface of femur smaller and present on only basal half of segment (extending past halfway in Jade-striped). Females indistinguishable.   **Habitat and Behavior:** Habitat preference similar to Jade-striped, overlaps geographically and can occur together. **Range:** Both slopes, to 600 m. Costa Rica to western Ecuador.

male

genus *Macrothemis* (p. 332)
selected male appendages in lateral and dorsal views

Deceptive Sylph
Macrothemis fallax (p. 340)

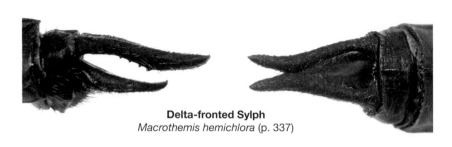

Delta-fronted Sylph
Macrothemis hemichlora (p. 337)

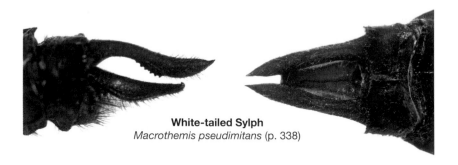

White-tailed Sylph
Macrothemis pseudimitans (p. 338)

TL 47 mm; HW 41 mm. **Identification:** Stream habitat, males with overall reddish color, orange wash on wings, and hindwings that are notably broad at base are good clues for this species. Abdomen with irregular black markings all over and less intensely red than most other red species such as tropical king skimmers, pondhawks, dragonlets, and meadowhawks. Females distinguished by moderate size and cryptically patterned black and tan thorax and abdomen. Perching on rocks (and sometimes ground) also distinctive. **Habitat and Behavior:** Inhabits small streams in forested landscape, though usually in sun. Males perch on rocks and fly up and down stream at intervals looking for females. Females, infrequently seen, also perch on rocks. Both sexes perch vertically on walls of bridges and houses, apparently treating them as rocks. Often feeds like flier, even associating with gliders. Uncommon. **Range:** Both slopes, 600–1300 m. Southwestern US to Panama.

male

female

TL 51 mm; HW 38 mm. **Identification:** Few dragonflies are plain brown to this degree, and no other has orange spot at hindwing nodus (and smaller patch at forewing nodus in some). Wing spots obvious in perched individuals, not usually visible in flight. Abdomen more tapered than in other similar-sized skimmers, becoming steadily narrow toward rear; distinctly more slender than somewhat similar Wandering Glider and lacking dark middorsal markings of glider. Long, down-curved cerci of females unique and good way to distinguish sexes. Dusk flight and hanging behavior also good identification clues, not duplicated by other brown species such as Tawny Pennant (p. 265) and Brown Setwing (p. 323). **Habitat and Behavior:** Breeds in open ponds and lakes of varying sizes; rarely breeds in pools in seasonal streams. Males appear in late afternoon and fly rapid and lengthy beats along shore and over water, interspersed with short periods of hovering. Observed mating at dusk followed by male guarding ovipositing female. Feeds at dusk and dawn like dusk-flying darners; comes out earlier than darners in evening and remains in flight later in morning. Hangs up at 45° angle on tree branches during day but also in more open situations than crepuscular darners, even in tall grass. **Range:** Both slopes, to 100 m; in migration sometimes to 1450 m. Texas, Florida, and West Indies to Chile.

male

female

HYACINTH GLIDERS genus *Miathyria*

Two species. The two species that occur in CR are the only members of their genus, the smallest among related fliers in the Libellulidae, with foraging habits much like their larger relatives. Their strong ties to floating vegetation at breeding sites as well as foraging in groups allies them behaviorally to pasture gliders, *Tauriphila*. Also compare with rainpool gliders (*Pantala*) and saddlebags (*Tramea*).

Hyacinth Glider *Miathyria marcella*

TL 39 mm; HW 31 mm. **Identification:** Distinguished by pale-striped brown or entirely purplish thorax; orangish, dark-tipped abdomen; and brown "saddles" on hindwings. When visible, stripes on sides of thorax separate it from other gliding dragonflies except Striped Saddlebags (p. 354), with similar golden wings and narrow saddle but larger and more robust, also redder when mature. Smaller Dwarf Glider is red, with large wing patch, very rarely as common at water or in feeding flocks as Hyacinth. **Habitat and Behavior:** Permanent and temporary wetlands, from small ponds and ditches to large lake margins, usually associated with water hyacinth or water lettuce, the larval habitat. Also seen over water lilies, even ovipositing females. Males fly back and forth over carpet of vegetation to watch for visiting females. Hang up to perch. Often seen in feeding flocks far from water, alone or with rainpool gliders, pasture gliders, and/or saddlebags, all of which have similar foraging behavior. Has been found in migrating groups in other regions. **Range:** Both slopes, to 800 m. Southern US to Argentina, Greater Antilles.

male

female

TL 33 mm; HW 27 mm. **Identification:** Small and red with prominent dark hindwing spot, superficially colored like Red-mantled Dragonlet but with totally different behavior, a flier rather than a percher. Distinguished from other species in foraging groups by small size. **Habitat and Behavior:** Ponds and temporary pools, usually associated with floating vegetation, both water lettuce and water hyacinth. Females oviposit in open water among vegetation mats. Hangs up to perch. Sometimes in feeding flocks over open areas, including with mixed-species groups. **Range:** Both slopes, to 100 m; migrants exceptionally to 1450 m. Northern Mexico to Peru and Brazil, Greater Antilles.

male

female

PASTURE GLIDERS genus *Tauriphila*

Three species. Pasture gliders breed in vegetation-covered ponds and feed aerially, often in swarms with other gliders and saddlebags. Between saddlebags and Hyacinth Gliders in size, they have similar flight style but smaller hindwing markings than either. The male abdomen is constricted slightly just beyond its base, but this is scarcely evident in females.

Arch-tipped Glider *Tauriphila argo*

TL 43 mm; HW 33 mm. **Identification:** Quite similar to Garnet Glider (p. 350) but lacks middorsal dark markings on S8-9, impossible to see except when perched. In CR, dark hindwing markings very slightly smaller. Check male appendages, relatively short and arched in Arch-tipped, slightly longer and straight in Garnet, visible with binoculars when they stop to hover. Females a bit duller and paler than males, probably indistinguishable from Garnet in field; in hand, Arch-tipped with at least two rows of doubled cells in radial planate, Garnet with one row. **Habitat and Behavior:** Marshy ponds in sun, males flying rapidly back and forth over territories 6–12 m in extent and periodically hovering, either over open water or above floating vegetation such as water lettuce. Feeds in swarms with other gliders and saddlebags. **Range:** Both slopes, to 800 m. Texas to Argentina, Cuba.

male

female

349

TL 45 mm; HW 37 mm. **Identification:** Male's straight appendages distinguish it from Arch-tipped Glider (p. 349), red abdomen from Aztec Glider. More contrast between bright red abdomen and very dark, even bluish, thorax than in Arch-tipped. Smaller size and small hindwing patch separate it from red species of saddlebags. Also considerably larger than Dwarf Glider. No other species show constricted abdomen base of male pasture gliders, although character not always evident in females. **Habitat and Behavior:** Marshy ponds in sun, males on territory usually associated with floating vegetation such as water lettuce and water hyacinth. Feeds in swarms with other gliders and saddlebags. **Range:** Both slopes, to 600 m. Central Mexico, Florida, and Greater Antilles to Bolivia and Brazil.

male

female

Aztec Glider *Tauriphila azteca*

TL 46 mm; HW 36 mm. **Identification:** Both sexes distinguished from two other pasture gliders by yellowish, dark-banded abdomen contrasting strongly with dark thorax. Banding varies from quite distinct to rather obscure. Overall color closer to Hyacinth Glider (p. 347) but larger and with smaller hindwing markings than that species. **Habitat and Behavior:** Marshy ponds in sun, males on territory associated with floating vegetation such as water lettuce. Feeds in swarms with other gliders and saddlebags. Much less common than other two pasture gliders in CR. **Range:** Northern Pacific slope, to 100 m. Northern Mexico and Florida to Costa Rica.

male

female

351

SADDLEBAGS   genus *Tramea*

Four species. Saddlebags are large skimmers. Like gliders, they have broad hindwings and spend a lot of time in flight. The males fly over water on territorial beats, and both sexes cruise over the landscape looking for feeding opportunities. They can be seen almost anywhere in the lowlands, often over open areas in small groups that could even be called swarms, mixed with rainpool and pasture gliders. Relatively large hindwing patches distinguish them from rainpool gliders and larger size from pasture gliders and Hyacinth Glider, but species of these groups can be difficult to distinguish in flight unless seen well. The immatures are paler and duller than mature individuals but share their basic field marks.

Saddlebags often perch horizontally at bare branch tips, even in the treetops, as well as hanging up more or less vertically, like gliders. Their mode of oviposition is unique. Males and females fly in tandem low over the water until they find an appropriate spot. Then the male releases the female, she drops to the surface and taps the water to lay a bunch of eggs, and the male picks her up again immediately to repeat the process elsewhere. The long, slender, and fairly simple cerci of the males surely facilitate this continued parting and reuniting. Less commonly, females oviposit alone.

Vermilion Saddlebags   *Tramea abdominalis*

TL 47 mm; HW 40 mm. **Identification:** Both sexes mostly red, with narrow wing patches ("saddles"). Females slightly duller than males, with thorax brown. Abdomen with black middorsal spots on S8-9. Male hamules protrude more in this and Red (p. 355) than in other two saddlebags, visible in silhouette. No hint of thoracic stripes as in Striped Saddlebags, and less black on abdomen tip than Striped. Somewhat like equally red Arch-tipped and Garnet Gliders, and may fly with them, but larger and with larger hindwing patches. **Habitat and Behavior:** Breeds in open ponds with scattered vegetation. **Range:** Both slopes, to 1200 m. Northern Mexico, Florida, and West Indies to Argentina.

male

female

TL 48 mm; HW 41 mm. **Identification:** Males are the only black saddlebags. Male cerci longer than in other saddlebags. Females and immature males most like female Vermilion Saddlebags, with reddish color and plain thorax, but larger black spots on S8–10 distinctive. No hint of pale stripes typical of Striped Saddlebags. **Habitat and Behavior:** Breeds in open ponds, usually fringed with vegetation. Males fly back and forth over water at head height, occasionally ascending to treetops and back down, and perch occasionally on tips of tall grasses. Not seen in mixed feeding swarms. **Range:** Both slopes, to 1000 m. Eastern Mexico and Greater Antilles to Argentina.

male

female

TL 47 mm; HW 41 mm. **Identification:** In females and younger males strong thoracic stripes are distinctive. Older males redder, with stripes obscured, distinguished from Vermilion Saddlebags (p. 352) by more black at abdomen tip (S8–10 mostly black in both sexes). Chance of mistaking this for Spot-winged Glider (p. 357) in flight because of vaguely striped look of thorax and hindwing spot of glider. Glider usually with more erratic and rapid flight. Note also possibility of confusion with pasture glider species that are equally red. In some lights, amber tint of wings of this species a good field mark. **Habitat and Behavior:** Breeds in open ponds, usually fringed with vegetation. Most common saddlebags, frequently in feeding swarms. Females often oviposit unattended. **Range:** Both slopes, to 800 m. Southern US to Argentina, West Indies.

male

female

TL 45 mm; HW 41 mm. **Identification:** Only saddlebags with broad hindwing patches. Among regional species, only a few dragonlets have equally large markings, and they are all smaller, fly less, and perch low in vegetation. **Habitat and Behavior:** Breeds in open ponds fringed with vegetation. Least common saddlebags in CR. **Range:** Both slopes, to 100 m. Southern US to Venezuela, Greater Antilles.

male

RAINPOOL GLIDERS genus *Pantala*

Two species. These fast, robust migrants have broad hindwings suited for gliding while flying long distances. They fly for extended periods to hunt, sometimes in feeding swarms, but are occasionally seen hanging up on a plant stem with abdomen vertical. Most breeding is in temporary, seasonal wetlands in open landscapes; they also breed in small pools, small, permanent bodies of water such as artificial ponds, and occasionally even in slow streams.

Wandering Glider *Pantala flavescens*

TL 48 mm; HW 40 mm. **Identification:** No other medium-sized skimmer is yellow-orange overall, nor has habit of long-term flying and then hanging up vertically. Other than Spot-winged Glider, could be mistaken for Evening Skimmer (p. 346), which is more slender and lacks black middorsal markings on abdomen. Females average a bit duller, especially on frons and abdomen, but often not distinguishable from males, as second-segment genitalia not clearly visible. Very close look at cerci might reveal epiproct below in males; female cerci slightly longer and straighter. **Habitat and Behavior:** Breeds in ponds, often seasonal, and isolated pools in seasonal streams. Males patrol over ponds and streams. Usually oviposits in tandem. Both sexes hover and dip over shiny automobiles, females even ovipositing on them. Otherwise can be seen in flight anywhere, even high above canopy. Hangs up in low herbaceous vegetation and higher on bare tree branches. Feeding swarms common over open country, rarely at dusk. Performs a seasonal east-west migration. **Range:** Both slopes, up to 1450 m during migration through Monteverde. Almost worldwide, mostly at tropical and subtropical latitudes but extending north in North America.

male

female

TL 48 mm; HW 40 mm. **Identification:** Readily distinguished from Wandering Glider when hind-wing spots visible, but they can be largely hidden by abdomen, so multiple views may be necessary. Spot-winged overall darker than Wandering, with more contrasty striped pattern on thorax and dis-tinctly redder head, at least in male. Could be confused with saddlebags, especially Striped (p. 354), unless hindwing pattern seen clearly. Saddlebags often perch horizontally on tips of twigs, gliders never, and male saddlebags often in territorial flights along shore of pond rather than over landscape. **Habitat and Behavior:** Breeds in temporary ponds and small streams. Typically oviposits solo but sometimes in tandem pairs. Migrates seasonally through Monteverde in August and early September, usually in small numbers, but sometimes hundreds or thousands per day. Southbound migratory flight seen in July over marshes at Los Chiles, most individuals at around 5 m in the air. Tends to hang up higher than Wandering, usually in trees. Swarm of more than 100 observed eating emerging fig wasps at crown of fruiting fig tree in September, near Siquirres. **Range:** Both slopes, to 1500 m. Southern Canada south to Argentina, West Indies, Galapagos.

male

female

Comparison of Color Patterns on Body of Rubyspots
(*Hetaerina* species) and Dancers (*Argia* species)

These two genera of damselflies present identification challenges; in the following pages, plates of scanned individuals are included to facilitate comparisons among the species of each genus.

Rubyspots (*Hetaerina* species): Males are presented opposite females for the nine rubyspot species.

Dancers (*Argia* species): Males are presented opposite females for the 28 dancer species. The species are placed within 7 groups based on similarities in coloration. Note that *Argia oculata* appears in both group 3 and group 5.

male *Hetaerina*

male
Hetaerina cruentata
(p. 50)

male
Hetaerina capitalis
(p. 51)

male
Hetaerina majuscula
(p. 52)

female *Hetaerina*

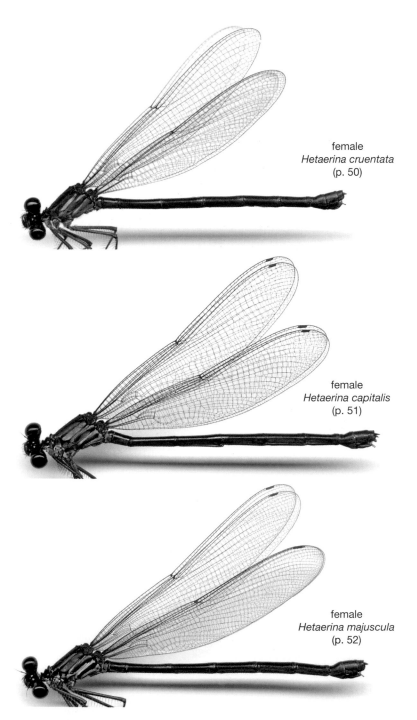

female
Hetaerina cruentata
(p. 50)

female
Hetaerina capitalis
(p. 51)

female
Hetaerina majuscula
(p. 52)

male *Hetaerina*

male
Hetaerina miniata
(p. 53)

male
Hetaerina caja
(p. 54)

male
Hetaerina occisa
(p. 55)

female *Hetaerina*

female
Hetaerina miniata
(p. 53)

female
Hetaerina caja
(p. 54)

female
Hetaerina occisa
(p. 55)

male *Hetaerina*

male
Hetaerina fuscoguttata
(p. 56)

male
Hetaerina sempronia
(p. 57)

male
Hetaerina titia
(p. 58)

female *Hetaerina*

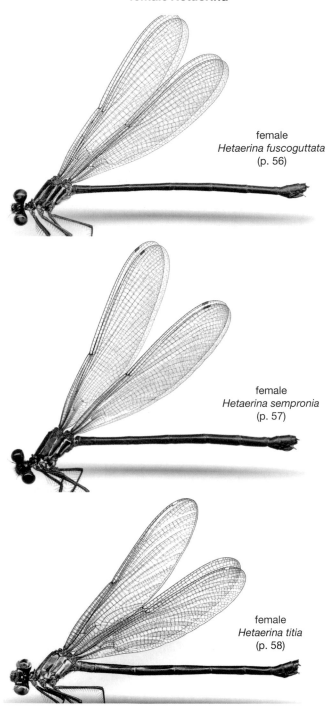

female
Hetaerina fuscoguttata
(p. 56)

female
Hetaerina sempronia
(p. 57)

female
Hetaerina titia
(p. 58)

Group 1
male *Argia*

male *Argia cupraurea*
(p. 81)

male *Argia oenea*
(p. 83)

male *Argia calverti*
(p. 84)

male *Argia fulgida*
(p. 85)

Group 1
female *Argia*

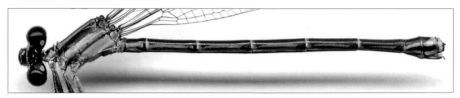

female *Argia cupraurea*
(p. 81)

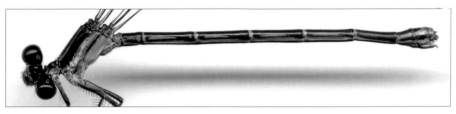

female *Argia oenea*
(p. 83)

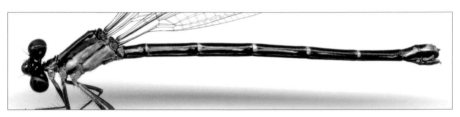

female *Argia calverti*
(p. 84)

female *Argia fulgida*
(p. 85)

Group 2
male *Argia*

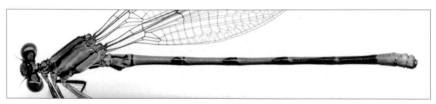

male *Argia anceps* (p. 86)

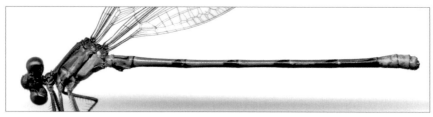

male *Argia fissa* (p. 87)

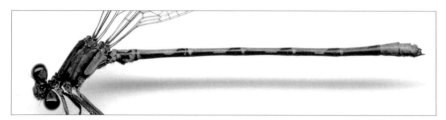

male *Argia elongata* (p. 88)

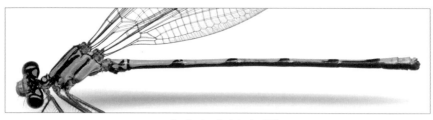

male *Argia chelata* (p. 89)

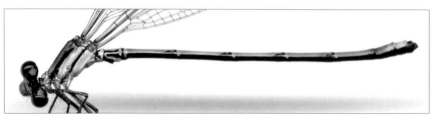

male *Argia medullaris* (p. 90)

Group 2
female *Argia*

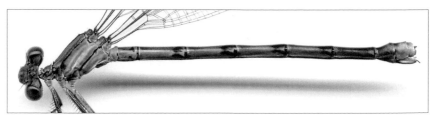

female *Argia anceps* (p. 86)

female *Argia fissa* (p. 87)

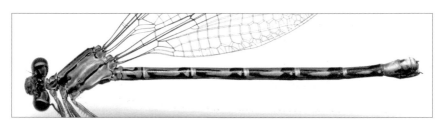

female *Argia elongata* (p. 88)

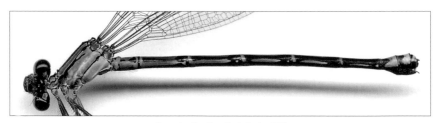

female *Argia chelata* (p. 89)

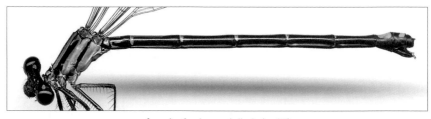

female *Argia medullaris* (p. 90)

Group 3
male *Argia*

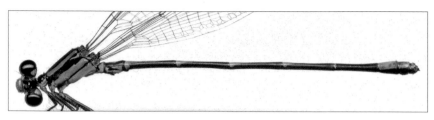

male *Argia oculata* (p. 91)
also placed in group 5

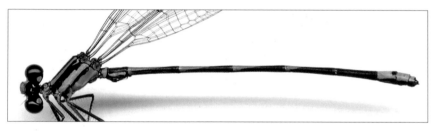

male *Argia adamsi* (p. 92)

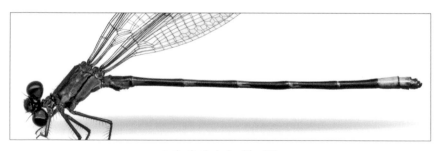

male *Argia haberi* (p. 94)

Group 3
female *Argia*

female *Argia oculata* (p. 91)

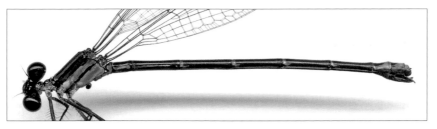

female *Argia adamsi* (p. 92)

female *Argia haberi* (p. 94)
specimen not available

Group 3
male *Argia*

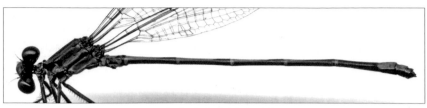

male *Argia insipida* (p. 95)

male *Argia popoluca* (p. 96), Pacific form

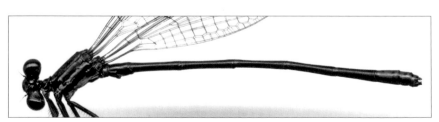

male *Argia popoluca* (p. 97), Caribbean form

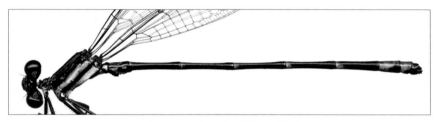

male *Argia schorri* (p. 98)

Group 3
female *Argia*

female *Argia insipida* (p. 95)
specimen not available

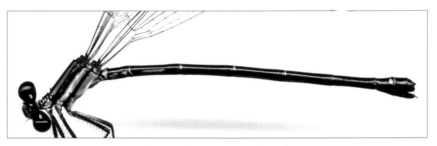

female *Argia popoluca* (p. 97)

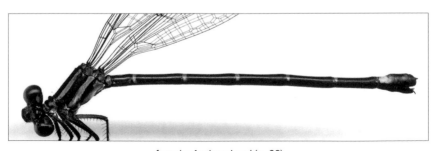

female *Argia schorri* (p. 98)

Group 3
male *Argia*

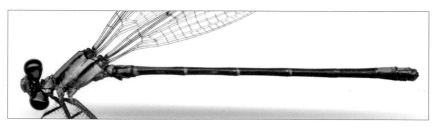

male *Argia talamanca* (p. 99)

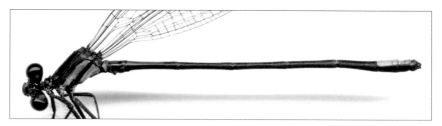

male *Argia underwoodi* (p. 100)

male *Argia terira* (p. 101)

Group 3
female *Argia*

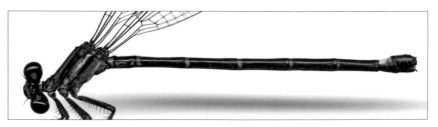

female *Argia talamanca* (p. 99)

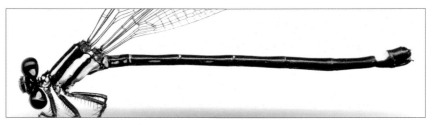

female *Argia underwoodi* (p. 100)

female *Argia terira* (p. 101)

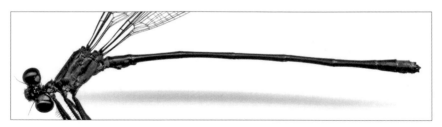

male *Argia carolus* (p. 102)

male *Argia indicatrix* (p. 103)

male *Argia rogersi* (p. 104)

Group 4
female *Argia*

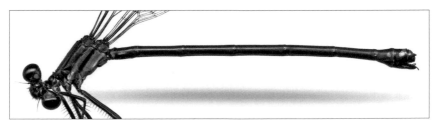

female *Argia carolus* (p. 102)

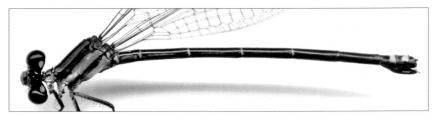

female *Argia indicatrix* (p. 103)

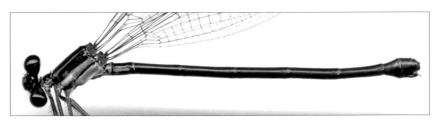

female *Argia rogersi* (p. 104)

Group 5
male *Argia*

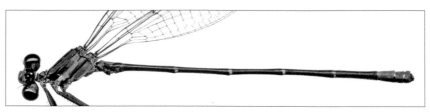

male *Argia frequentula* (p. 106)

male *Argia pulla* (p. 107)

male *Argia johannella* (p. 108)

male *Argia ulmeca* (p. 109)

male *Argia oculata* (p. 91)

Group 5
female *Argia*

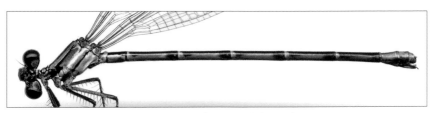

female *Argia frequentula* (p. 106)

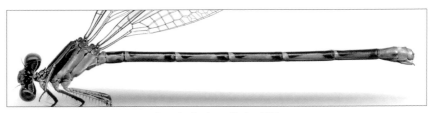

female *Argia pulla* (p. 107)

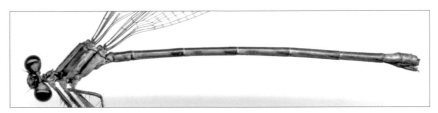

female *Argia johannella* (p. 108)

female *Argia ulmeca* (p. 109)

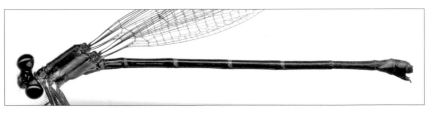

female *Argia oculata* (p. 91)

Group 6
male *Argia*

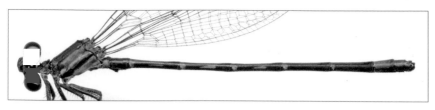

male *Argia pocomana* (p. 110)

Group 7
male *Argia*

male *Argia translata* (p. 111)

male *Argia tezpi* (p. 113)

Group 6
female *Argia*

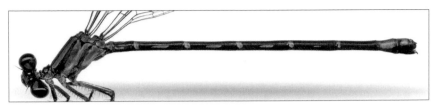

female *Argia pocomana* (p. 110)

Group 7
female *Argia*

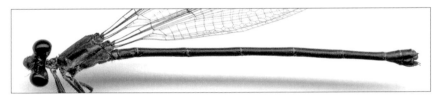

female *Argia translata* (p. 111)

female *Argia tezpi* (p. 113)

Appendix A
Species Previously Recorded from Costa Rica

As this book also stands as the formal list of the Odonata of Costa Rica, the authors here include a list of scientific names that have been reported from the country in the literature but are not included in the book. Some have been combined with other species, while others have been found to be erroneously identified or merely reported in error.

Anatya normalis combined with *A. guttata.*
Aphylla obscura combined with *A. tenuis.*
Argia cuprea reidentified as *A. fulgida.*
Argia difficilis reidentified as *A. oculata.*
Argia eliptica synonym of *A. oculata.*
Argia extranea split into two species; CR one is *A. elongata.*
Argia gaumeri listed erroneously.
Argia variabilis reidentified as *A. medullaris.*
Chrysobasis lucifer now *Leptobasis lucifer.*
Cora chirripa now *Miocora chirripa.*
Cora notoxantha combined with *Miocora semiopaca.*
Cora obscura combined with *Miocora semiopaca.*
Cora semiopaca now *Miocora semiopaca.*
Cora skinneri combined with *Miocora chirripa.*
Cordulegaster godmani combined with *C. diadema.*
Dythemis multipunctata renamed *D. nigra.*
Erythrodiplax famula reidentified as *E. lativittata.*
Hetaerina pilula dropped from list; identity of specimen dubious.
Heteragrion atrolineatum combined with *H. calendulum.*
Heteragrion atroterminatum combined with *H. mitratum.*
Idiataphe amazonica record could not be verified.
Lestes scalaris reidentified as *L. tikalus.*
Macrodiplax balteata listed erroneously.
Macrothemis tessellata inequiunguis now considered a full species.
Metaleptobasis westfalli combined with *M. foreli.*
Micrathyria tibialis listed erroneously.
Nephepeltia chalconota combined with *N. flavifrons.*
Perithemis mooma combined with *P. tenera.*
Philogenia lankesteri described from a single female, not considered a valid species.
Psaironeura remissa reidentified as *P. angeloi.*
Pseudostigma accedens listed in literature, but no valid specimen seen.
Sympetrum corruptum listed erroneously.
Telebasis griffinii reidentified as *T. levis.*
Tramea insularis listed erroneously.
Triacanthagyna dentata listed erroneously.

Appendix B
List of Costa Rican Species
(only species described in literature)

TWIGTAILS, PERILESTIDAE
Horned Twigtail, *Perissolestes magdalenae* (Williamson & Williamson, 1924)
Green-striped Twigtail, *Perissolestes remotus* (Williamson & Williamson, 1924)

SPREADWINGS, LESTIDAE
Great Spreadwing, *Archilestes grandis* (Rambur, 1842)
Yellow-eyed Spreadwing, *Archilestes latialatus* Donnelly, 1982
Cloudforest Spreadwing, *Archilestes neblina* Garrison, 1982
Plateau Spreadwing, *Lestes alacer* Hagen, 1861
Rainpool Spreadwing, *Lestes forficula* Rambur, 1842
Montane Spreadwing, *Lestes henshawi* Calvert, 1907
Chalky Spreadwing, *Lestes sigma* Calvert, 1901
Blue-striped Spreadwing, *Lestes tenuatus* Rambur, 1842
Tikal Spreadwing, *Lestes tikalus* Kormondy, 1959

SHADOWDAMSELS, PLATYSTICTIDAE
Cacao Shadowdamsel, *Palaemnema baltodanoi* Brooks, 1989
Chiriquita Shadowdamsel, *Palaemnema chiriquita* Calvert, 1931
Arch-tipped Shadowdamsel, *Palaemnema collaris* Donnelly, 1992
Dentate Shadowdamsel, *Palaemnema dentata* Donnelly, 1992
Carrillo Shadowdamsel, *Palaemnema distadens* Calvert, 1931
Elongate Shadowdamsel, *Palaemnema gigantula* Calvert, 1931
Janet's Shadowdamsel, *Palaemnema joanetta* Kennedy, 1940
Black-backed Shadowdamsel, *Palaemnema melanota* Ris, 1918
Nathalia Shadowdamsel, *Palaemnema nathalia* Selys, 1886
Ink-tipped Shadowdamsel, *Palaemnema paulirica* Calvert, 1931
Reventazon Shadowdamsel, *Palaemnema reventazoni* Calvert, 1931

BROAD-WINGED DAMSELS, CALOPTERYGIDAE
River Rubyspot, *Hetaerina caja* (Drury, 1773)
Bronze Rubyspot, *Hetaerina capitalis* Selys, 1873
Highland Rubyspot, *Hetaerina cruentata* (Rambur, 1842)
Dot-winged Rubyspot, *Hetaerina fuscoguttata* Selys, 1878
Purplish Rubyspot, *Hetaerina majuscula* Selys, 1853
Red-striped Rubyspot, *Hetaerina miniata* Selys, 1879
Racket-tipped Rubyspot, *Hetaerina occisa* Hagen in Selys, 1853
Forest Rubyspot, *Hetaerina sempronia* Hagen in Selys, 1853
Smoky Rubyspot, *Hetaerina titia* (Drury, 1773)

BANNERWINGS, POLYTHORIDAE
Blue Cora, *Cora marina* Selys, 1868
Chirripo Cora, *Miocora chirripa* (Calvert, 1907)
Peralta Cora, *Miocora peraltica* Calvert, 1917
Variable Cora, *Miocora semiopaca* (Selys, 1878)

FLAMBOYANT FLATWINGS, HETERAGRIONIDAE
Pale-faced Flatwing, *Heteragrion albifrons* Ris, 1918
Golden-faced Flatwing, *Heteragrion calendulum* Williamson, 1919
Red-and-black Flatwing, *Heteragrion erythrogastrum* Selys, 1886

Mountain Flatwing, *Heteragrion majus* Selys, 1886
Orange-banded Flatwing, *Heteragrion mitratum* Williamson, 1919

DUSKY FLATWINGS, PHILOGENIIDAE
Costa Rican Flatwing, *Philogenia carrillica* Calvert, 1907
Golfo Dulce Flatwing, *Philogenia championi* Calvert, 1901
Limon Flatwing, *Philogenia expansa* Calvert, 1924
Blue Flatwing, *Philogenia peacocki* Brooks, 1989
Terraba Flatwing, *Philogenia terraba* Calvert, 1907

CASCADE DAMSELS, THAUMATONEURIDAE
Great Cascade Damsel, *Thaumatoneura inopinata* McLachlan, 1897

POND DAMSELS, COENAGRIONIDAE
Narrow-tipped Wedgetail, *Acanthagrion inexpectum* Leonard, 1977
Costa Rican Wedgetail, *Acanthagrion speculum* Garrison, 1985
Pacific Wedgetail, *Acanthagrion trilobatum* Leonard, 1977
Middle American Pearlwing, *Anisagrion allopterum* Selys, 1876
Kennedy's Pearlwing, *Anisagrion kennedyi* Leonard, 1937
Varied Dancer, *Argia adamsi* Calvert, 1902
Cerulean Dancer, *Argia anceps* Garrison, 1996
Calvert's Dancer, *Argia calverti* Garrison & von Ellenrieder, 2017
Blue-gray Dancer, *Argia carolus* Garrison & von Ellenrieder, 2017
Big Blue Dancer, *Argia chelata* Calvert, 1902
Ruby Dancer, *Argia cupraurea* Calvert, 1902
Thorn-tipped Dancer, *Argia elongata* Garrison & von Ellenrieder, 2017
Azure Dancer, *Argia fissa* Selys, 1865
Green-eyed Dancer, *Argia frequentula* Calvert, 1907
Garnet Dancer, *Argia fulgida* Navás, 1934
Mountain Spring Dancer, *Argia haberi* Garrison & von Ellenrieder, 2017
Swamp Dancer, *Argia indicatrix* Calvert, 1902
River Dancer, *Argia insipida* Hagen in Selys, 1865
Bristle-tipped Dancer, *Argia johannella* Calvert, 1907
Sky-blue Dancer, *Argia medullaris* Hagen in Selys, 1865
Oculate Dancer, *Argia oculata* Hagen in Selys, 1865
Fiery-eyed Dancer, *Argia oenea* Hagen in Selys, 1865
Pocomana Dancer, *Argia pocomana* Calvert, 1907
Popoluca Dancer, *Argia popoluca* Calvert, 1902
Purple Dancer, *Argia pulla* Hagen in Selys, 1865
Black-fronted Dancer, *Argia rogersi* Calvert, 1902
Golfo Dulce Dancer, *Argia schorri* Garrison & von Ellenrieder, 2017
Talamanca Dancer, *Argia talamanca* Calvert, 1907
Terira Dancer, *Argia terira* Calvert, 1907
Tezpi Dancer, *Argia tezpi* Calvert, 1902
Dusky Dancer, *Argia translata* Hagen in Selys, 1865
Olmec Dancer, *Argia ulmeca* Calvert, 1902
Waterfall Dancer, *Argia underwoodi* Calvert, 1907
Familiar Bluet, *Enallagma civile* (Hagen, 1861)
Neotropical Bluet, *Enallagma novaehispaniae* Calvert, 1907
Tiny Forktail, *Ischnura capreolus* Hagen, 1861
Citrine Forktail, *Ischnura hastata* (Say, 1840)
Rambur's Forktail, *Ischnura ramburii* (Selys, 1850)
Guanacaste Swampdamsel, *Leptobasis guanacaste* Paulson, 2009
Lucifer Swampdamsel, *Leptobasis lucifer* (Donnelly, 1967)
Red-tipped Swampdamsel, *Leptobasis vacillans* Hagen in Selys, 1877

Long-tailed Helicopter, *Mecistogaster linearis* (Fabricius, 1776)
Bromeliad Helicopter, *Mecistogaster modesta* Selys, 1860
Lemon-tipped Helicopter, *Mecistogaster ornata* Rambur, 1842
Blue-winged Helicopter, *Megaloprepus caerulatus* (Drury, 1782)
Guatemalan Spinyneck, *Metaleptobasis bovilla* Calvert, 1907
Panamanian Spinyneck, *Metaleptobasis foreli* Ris, 1918
Tropical Sprite, *Nehalennia minuta* (Selys, 1857)
Caribbean Yellowface, *Neoerythromma cultellatum* (Selys, 1876)
Amelia's Threadtail, *Neoneura amelia* Calvert, 1903
Esther's Threadtail, *Neoneura esthera* Williamson, 1917
Crimson Threadtail, *Protoneura amatoria* Calvert, 1907
Golden-orange Threadtail, *Protoneura aurantiaca* Selys, 1886
Sulfury Threadtail, *Protoneura sulfurata* Donnelly, 1989
Wispy Threadtail, *Psaironeura angeloi* Tennessen, 2016
Selva Threadtail, *Psaironeura selvatica* Esquivel, 1993
Broad-winged Helicopter, *Pseudostigma aberrans* Selys, 1860
Golden Firetail, *Telebasis aurea* May, 1992
Belize Firetail, *Telebasis boomsmae* Garrison, 1994
Coral Firetail, *Telebasis corallina* (Selys, 1876)
Marsh Firetail, *Telebasis digiticollis* Calvert, 1902
Striped Firetail, *Telebasis filiola* (Perty, 1833)
Montane Firetail, *Telebasis garleppi* Ris, 1918
Green-eyed Firetail, *Telebasis isthmica* Calvert, 1902
Hyacinth Firetail, *Telebasis levis* Garrison, 2009
Red-and-black Firetail, *Telebasis rojinegra* Haber, 2020
Desert Firetail, *Telebasis salva* (Hagen, 1861)

DARNERS, AESHNIDAE
Mountain Stream Darner, *Aeshna williamsoniana* Calvert, 1905
Amazon Darner, *Anax amazili* (Burmeister, 1839)
Blue-spotted Comet Darner, *Anax concolor* Brauer, 1865
Blue-faced Darner, *Coryphaeschna adnexa* (Hagen, 1861)
Amazon Red Darner, *Coryphaeschna amazonica* De Marmels, 1989
Icarus Darner, *Coryphaeschna apeora* Paulson, 1994
Fiery Darner, *Coryphaeschna diapyra* Paulson, 1994
Mangrove Darner, *Coryphaeschna viriditas* Calvert, 1952
Auricled Darner, *Gynacantha auricularis* Martin, 1909
Yellow-legged Darner, *Gynacantha caudata* Karsch, 1891
Pale-banded Darner, *Gynacantha gracilis* (Burmeister, 1839)
Little Brown Darner, *Gynacantha laticeps* Williamson, 1923
Dark-saddled Darner, *Gynacantha membranalis* Karsch, 1891
Bar-sided Darner, *Gynacantha mexicana* Selys, 1868
Twilight Darner, *Gynacantha nervosa* Rambur, 1842
Gold-tipped Darner, *Gynacantha tibiata* Karsch, 1891
Chartreuse Darner, *Gynacantha vargasi* Haber, 2019
Mayan Evening Darner, *Neuraeschna maya* Belle, 1989
Malachite Darner, *Remartinia luteipennis* (Burmeister, 1839)
Highland Darner, *Rhionaeschna cornigera* (Brauer, 1865)
Black-tailed Darner, *Rhionaeschna jalapensis* (Williamson, 1908)
Turquoise-tipped Darner, *Rhionaeschna psilus* (Calvert, 1947)
Magnificent Megadarner, *Staurophlebia reticulata* (Burmeister, 1839)
Caribbean Darner, *Triacanthagyna caribbea* Williamson, 1923
Ditzler's Darner, *Triacanthagyna ditzleri* Williamson, 1923
Satyr Darner, *Triacanthagyna satyrus* (Martin, 1909)
Pale-green Darner, *Triacanthagyna septima* (Selys, 1857)

CLUBTAILS, GOMPHIDAE

Slender Clubtail, *Agriogomphus tumens* (Calvert, 1905)
Broad-striped Forceptail, *Aphylla angustifolia* Garrison, 1986
Narrow-striped Forceptail, *Aphylla protracta* (Hagen in Selys, 1859)
Obscure Forceptail, *Aphylla tenuis* Selys, 1859
Mexican Snout-tail, *Archaeogomphus furcatus* Williamson, 1923
Pincertail, Neotropical, *Desmogomphus paucinervis* (Selys, 1873)
Armed Knobtail, *Epigomphus armatus* Ris, 1918
Cloudforest Knobtail, *Epigomphus bosquenuboso* Haber, 2017
Camel Knobtail, *Epigomphus camelus* Calvert, 1905
Horned Knobtail, *Epigomphus corniculatus* Belle, 1989
Plate-crowned Knobtail, *Epigomphus echeverrii* Brooks, 1989
Limon Knobtail, *Epigomphus houghtoni* Brooks, 1989
Morrison's Knobtail, *Epigomphus morrisoni* Haber, 2017
Fork-tipped Knobtail, *Epigomphus quadracies* Calvert, 1903
Common Knobtail, *Epigomphus subobtusus* Selys, 1878
Hook-tipped Knobtail, *Epigomphus subsimilis* Calvert, 1920
Lowland Knobtail, *Epigomphus tumefactus* Calvert, 1903
Cartago Knobtail, *Epigomphus verticicornis* Calvert, 1908
Wagner's Knobtail, *Epigomphus wagneri* Haber, 2017
One-striped Ringtail, *Erpetogomphus bothrops* Garrison, 1994
Knob-tipped Ringtail, *Erpetogomphus constrictor* Ris, 1918
Lime Ringtail, *Erpetogomphus elaphe* Garrison, 1994
Blue-faced Ringtail, *Erpetogomphus eutainia* Calvert, 1905
Schaus's Ringtail, *Erpetogomphus schausi* Calvert, 1919
Tristan's Ringtail, *Erpetogomphus tristani* Calvert, 1912
Isthmian Pegtail, *Perigomphus pallidistylus* (Belle, 1972)
Ringed Forceptail, *Phyllocycla breviphylla* Belle, 1975
Tiny Forceptail, *Phyllocycla volsella* (Calvert, 1905)
Panamanian Leaftail, *Phyllogomphoides appendiculatus* (Kirby, 1899)
Two-striped Leaftail, *Phyllogomphoides bifasciatus* (Hagen in Selys, 1878)
Horned Leaftail, *Phyllogomphoides burgosi* Brooks, 1989
Tuxtla Leaftail, *Phyllogomphoides pugnifer* Donnelly, 1979
Common Leaftail, *Phyllogomphoides suasus* (Selys, 1859)
Anomalous Sanddragon, *Progomphus anomalus* Belle, 1973
Zebra-striped Sanddragon, *Progomphus clendoni* Calvert, 1905
Forest Sanddragon, *Progomphus longistigma* Ris, 1918
Mexican Sanddragon, *Progomphus mexicanus* Belle, 1973
Pygmy Sanddragon, *Progomphus pygmaeus* Selys, 1873

SPIKETAILS, CORDULEGASTRIDAE

Apache Spiketail, *Cordulegaster diadema* Selys, 1868

EMERALDS, CORDULIIDAE (perhaps not in this family)

Bates's Emerald, *Neocordulia batesi* (Selys, 1871)
Cerro Campana Emerald, *Neocordulia campana* May & Knopf, 1988
Elusive Emerald, *Neocordulia griphus* May, 1992

SKIMMERS, LIBELLULIDAE

Common Blue-eye, *Anatya guttata* (Erichson in Schomburgk, 1848)
Red-tailed Pennant, *Brachymesia furcata* (Hagen, 1861)
Tawny Pennant, *Brachymesia herbida* (Gundlach, 1889)
Little Clubskimmer, *Brechmorhoga nubecula* (Rambur, 1842)
Masked Clubskimmer, *Brechmorhoga pertinax* (Hagen, 1861)
Slender Clubskimmer, *Brechmorhoga praecox* (Hagen, 1861)

Amber-banded Clubskimmer, *Brechmorhoga rapax* Calvert, 1898
Vivacious Clubskimmer, *Brechmorhoga vivax* Calvert, 1906
Gray-waisted Skimmer, *Cannaphila insularis* Kirby, 1889
Morton's Skimmer, *Cannaphila mortoni* Donnelly, 1992
Blue-tailed Skimmer, *Cannaphila vibex* (Hagen, 1861)
Blue-eyed Setwing, *Dythemis nigra* Martin, 1897
Brown Setwing, *Dythemis sterilis* Hagen, 1861
Apricot Streamskimmer, *Elasmothemis aliciae* González-Soriano & Novelo-Gutiérrez, 2006
Ruddy Streamskimmer, *Elasmothemis cannacrioides* (Calvert, 1906)
Fairy Skimmer, *Elga leptostyla* Ris, 1909
Black Pondhawk, *Erythemis attala* (Selys in Sagra, 1857)
Little Pondhawk, *Erythemis credula* (Hagen, 1861)
Red Pondhawk, *Erythemis haematogastra* (Burmeister, 1839)
Claret Pondhawk, *Erythemis mithroides* (Brauer, 1900)
Flame-tailed Pondhawk, *Erythemis peruviana* (Rambur, 1842)
Pin-tailed Pondhawk, *Erythemis plebeja* (Burmeister, 1839)
Eastern Pondhawk, *Erythemis simplicicollis* (Say, 1840)
Great Pondhawk, *Erythemis vesiculosa* (Fabricius, 1775)
Montane Dragonlet, *Erythrodiplax abjecta* (Rambur, 1842)
Andagoya Dragonlet, *Erythrodiplax andagoya* Borror, 1942
Seaside Dragonlet, *Erythrodiplax berenice* (Drury, 1773)
Scarlet Dragonlet, *Erythrodiplax castanea* (Burmeister, 1839)
Red-mantled Dragonlet, *Erythrodiplax fervida* (Erichson in Schomburgk, 1848)
Black-winged Dragonlet, *Erythrodiplax funerea* (Hagen, 1861)
Red-faced Dragonlet, *Erythrodiplax fusca* (Rambur, 1842)
Chalk-marked Dragonlet, *Erythrodiplax kimminsi* Borror, 1942
Canopy Dragonlet, *Erythrodiplax laselva* Haber, Wagner & de la Rosa, 2015
Coffee Bean Dragonlet, *Erythrodiplax lativittata* Borror, 1942
Band-winged Dragonlet, *Erythrodiplax umbrata* (Linnaeus, 1758)
Metallic Pennant, *Idiataphe cubensis* (Scudder, 1866)
Neon Skimmer, *Libellula croceipennis* Selys, 1868
Highland Skimmer, *Libellula foliata* (Kirby, 1889)
Hercules Skimmer, *Libellula herculea* Karsch, 1889
Maria's Skimmer, *Libellula mariae* Garrison, 1992
Gold-spotted Sylph, *Macrothemis aurimaculata* Donnelly, 1984
Delia Sylph, *Macrothemis delia* Ris, 1913
Attenuate Sylph, *Macrothemis extensa* Ris, 1913
Deceptive Sylph, *Macrothemis fallax* May, 1998
Delta-fronted Sylph, *Macrothemis hemichlora* (Burmeister, 1839)
Ivory-striped Sylph, *Macrothemis imitans* Karsch, 1890
Straw-colored Sylph, *Macrothemis inacuta* Calvert, 1898
Jade-striped Sylph, *Macrothemis inequiunguis* Calvert, 1895
Delicate Sylph, *Macrothemis musiva* Calvert, 1898
White-tailed Sylph, *Macrothemis pseudimitans* Calvert, 1898
Hyacinth Glider, *Miathyria marcella* (Selys in Sagra, 1857)
Dwarf Glider, *Miathyria simplex* (Rambur, 1842)
Spot-tailed Dasher, *Micrathyria aequalis* (Hagen, 1861)
Black Dasher, *Micrathyria atra* (Martin, 1897)
Dark-fronted Dasher, *Micrathyria catenata* Calvert, 1909
Peten Dasher, *Micrathyria debilis* (Hagen, 1861)
Even-striped Dasher, *Micrathyria dictynna* Ris, 1919
Three-striped Dasher, *Micrathyria didyma* (Selys in Sagra, 1857)
Thornbush Dasher, *Micrathyria hagenii* Kirby, 1890
Swamp Dasher, *Micrathyria laevigata* Calvert, 1909
Fork-tipped Dasher, *Micrathyria mengeri* Ris, 1919

Square-spotted Dasher, *Micrathyria ocellata* Martin, 1897
Little Swamp Dasher, *Micrathyria pseudeximia* Westfall, 1992
Dusky Dasher, *Micrathyria schumanni* Calvert, 1906
Forest Pond Dasher, *Micrathyria venezuelae* De Marmels, 1989
Glossy-fronted Dryad, *Nephepeltia flavifrons* (Karsch, 1889)
Spine-bellied Dryad, *Nephepeltia phryne* (Perty, 1833)
Sunshine Leafsitter, *Oligoclada heliophila* Borror, 1931
Shadowy Leafsitter, *Oligoclada umbricola* Borror, 1931
Side-striped Skimmer, *Orthemis aequilibris* Calvert, 1909
Yellow-lined Skimmer, *Orthemis biolleyi* Calvert, 1906
Swamp Skimmer, *Orthemis cultriformis* Calvert, 1899
Carmine Skimmer, *Orthemis discolor* (Burmeister, 1839)
Roseate Skimmer, *Orthemis ferruginea* (Fabricius, 1775)
Slender Skimmer, *Orthemis levis* Calvert, 1906
Red-tailed Skimmer, *Orthemis schmidti* Buchholz, 1950
Red Rock Skimmer, *Paltothemis lineatipes* Karsch, 1890
Wandering Glider, *Pantala flavescens* (Fabricius, 1798)
Spot-winged Glider, *Pantala hymenaea* (Say, 1840)
Slough Amberwing, *Perithemis domitia* (Drury, 1773)
Golden Amberwing, *Perithemis electra* Ris, 1930
Eastern Amberwing, *Perithemis tenera* Say, 1840
Mexican Scarlet-tail, *Planiplax sanguiniventris* (Calvert, 1907)
Filigree Skimmer, *Pseudoleon superbus* (Hagen, 1861)
Brilliant Redskimmer, *Rhodopygia hinei* Calvert, 1907
Cardinal Meadowhawk, *Sympetrum illotum* (Hagen, 1861)
Talamanca Meadowhawk, *Sympetrum nigrocreatum* Calvert, 1920
Arch-tipped Glider, *Tauriphila argo* (Hagen, 1869)
Garnet Glider, *Tauriphila australis* (Hagen, 1867)
Aztec Glider, *Tauriphila azteca* Calvert, 1906
Evening Skimmer, *Tholymis citrina* Hagen, 1867
Vermilion Saddlebags, *Tramea abdominalis* (Rambur, 1842)
Sooty Saddlebags, *Tramea binotata* (Rambur, 1842)
Striped Saddlebags, *Tramea calverti* Muttkowski, 1910
Red Saddlebags, *Tramea onusta* Hagen, 1861
Large Woodskimmer, *Uracis fastigiata* (Burmeister, 1839)
Tropical Woodskimmer, *Uracis imbuta* (Burmeister, 1839)
Turrialba Woodskimmer, *Uracis turrialba* Ris, 1919
Amazon Sapphirewing, *Zenithoptera fasciata* (Linnaeus, 1758)

Appendix C
List of Species Known from Adjacent Countries
(but not reported for Costa Rica)

Nicaragua

Angelina's Shadowdamsel, *Palaemnema angelina*
Desert Shadowdamsel, *Palaemnema domina*
Honduran Shadowdamsel, *Palaemnema paulina*
American Rubyspot, *Hetaerina americana*
Ivory-faced Flatwing, *Heteragrion eboratum*
Mexican Wedgetail, *Acanthagrion quadratum*
Westfall's Knobtail, *Epigomphus westfalli*
Double-toothed Leaftail, *Phyllogomphoides duodentatus*

Panama

Scythe-tipped Spreadwing, *Lestes secula*
Spot-fronted Shadowdamsel, *Palaemnema bilobulata*
Hook-tipped Shadowdamsel, *Palaemnema cyclohamulata*
Black-tailed Shadowdamsel, *Palaemnema melanura*
Coppery Shadowdamsel, *Palaemnema mutans*
Spinulate Shadowdamsel, *Palaemnema spinulata*
Red-faced Flatwing, *Heteragrion rubrifulvum*
Black-and-yellow Flatwing, *Heteragrion valgum*
August's Flatwing, *Philogenia augusti*
Leonora's Flatwing, *Philogenia leonora*
Zetek's Flatwing, *Philogenia zeteki*
Superb Redleg, *Heteropodagrion superbum*
Kennedy's Wedgetail, *Acanthagrion kennedii*
Letitia Threadtail, *Drepanoneura letitia*
Lemon-striped Threadtail, *Neoneura confundens*
Wide-winged Helicopter, *Pseudostigma accedens* (also to the north)
Swamp Firetail, *Telebasis griffinii*
Williamson's Firetail, *Telebasis williamsoni*
Jesse's Darner, *Gynacantha jessei*
Black-legged Darner, *Triacanthagyna dentata*
One-striped Darner, *Triacanthagyna obscuripennis*
Blunt-tipped Knobtail, *Epigomphus compactus*
Cerro Campana Knobtail, *Epigomphus subquadrices*
Colombian Ringtail, *Erpetogomphus sabaleticus*
Dark-shouldered Leaftail, *Phyllogomphoides insignatus*
Side-spotted Leaftail, *Phyllogomphoides litoralis*
Carmelita Pondhawk, *Erythemis carmelita*
White-tailed Dragonlet, *Erythrodiplax unimaculata*
Amazon Pennant, *Idiataphe amazonica* (also to the north)
Pegtooth Sylph, *Macrothemis nobilis*
Blue-tipped Dasher, *Micrathyria caerulistyla*
Pale-legged Dasher, *Micrathyria tibialis*
Stripe-fronted Dryad, *Nephepeltia leonardina*
Reddish-brown Skimmer, *Orthemis aciculata*
Garrison's Skimmer, *Orthemis garrisoni*
Guiana Scarlet-tail, *Planiplax phoenicura*

Glossary

anal loop. Set of cells near the base of the hindwings of emerald and skimmer dragonflies that is distinctively shaped and strengthens the wing base.

andromorph. Male-like morph in odonates in which females exhibit two coloration morphs.

anteclypeus. Segment of lower part of face above labrum.

antehumeral stripe. One of pair on sides of thorax, bordered above by median stripe and below by humeral stripe.

antenodal cells. Cells between quadrangle and nodus in damselflies.

antenodal crossveins. Crossveins at fore edge of wing between base and nodus.

anterior lamina. Ventrolateral part of segment bordering front of male genital fossa on either side, sometimes expanded.

apical. *adj*. Near or at the tip of a structure.

auricle. A projection from either side of segment 2, most prominent in male darners.

banded. *adj*. With markings that run across the body or wing axis.

basal. *adj*. Near or at the base of a structure.

bridge crossveins. Crossveins running through triangular cell behind nodus.

carina. Raised ridge between two body segments.

caudal lamellae. Three gills or gill-like structures at tip of abdomen in larval damselflies.

cercus (pl. cerci). One of the superior terminal appendages in both male and female odonates.

costa. The vein forming the anterior edge of the wing.

coxa. Basal segment of leg.

crepuscular. *adj*. Active at dusk and dawn.

cuticle. Outer covering of an insect.

denticles. Very small teeth, often on cerci or abdominal segments.

diapause. A state of dormancy.

distal. *adj*. Toward tip or end.

dorsal. *adj*. The top or upperparts.

dorsolateral. *adj*. Where top and side meet; dorsolateral markings can extend well onto top and/ or side.

epiproct. Lower terminal appendage of male anisopterans.

exuvia (pl. exuviae). Shell-like cuticle left behind by an emerging adult odonate or a molting larva.

face. Front part of head, consisting of frons, postclypeus, anteclypeus, and labrum, from top to bottom.

femur (pl. femora). Long second segment of leg between coxa and tibia.

forcipate. *adj*. Forcepslike.

frons. Upper part and upper surface of face.

front. Of thorax, part between prothorax and wings; also called "top" in text, as it seems to be the top in damselflies.

genital fossa. Hollow on underside of male segment 2 in which genitalia lie.

genital lobe. Lobe at rear end of male segment 2 on either side of genital fossa.

genital pore. Pore under eighth abdominal segment through which sperm (from male) and eggs (from female) are excreted.

hamules (also hamular processes). Paired structures in male genital fossa that grasp female abdomen during copulation, often distinctive of species.

heteromorph. Morph differing from male in odonates in which females exhibit two color morphs.

humeral stripe. One of pair of dark stripes on sides of thorax bordered above by antehumeral stripe and below by pale color.

labium. "Lower lip," broad mouthpart under mandibles; much larger in larvae.

labrum. "Upper lip," lowest part of face.

lateral. *adj*. On side.

lek. Gathering of males for group display to attract females.

mandibles. Chewing mouthparts between labrum and labium.

median. *adj.* In middle, down midline.

medial planate. Distinctively shaped row of cells in middle of wing.

medio-dorsal. *adj.* Viewed from above and inward.

mesostigmal lamina (pl. laminae; also mesostigmal plate). One of paired transverse plates at very front of pterothorax, often used to distinguish females of similar damselfly species.

montane. *adj.* In mountains.

nodus. Prominent strengthening crossvein at front of wing, near middle.

obelisk. Position some dragonflies assume with abdomen highly elevated, even vertical, pointing toward sun to minimize heating.

occipital bar. Pale bar, often interrupted, between postocular spots in some damselflies.

occiput. Back of the head, visible as a small plate between the rear part of the eyes on dragonflies.

ocellar groove. Groove between ocelli.

ocellus (pl. ocelli). One of three light-gathering "simple eyes" on top of head.

ovipositor. Structure at end of abdomen of damselflies and some dragonflies for laying eggs.

paraproct. One of paired lower terminal appendages of male zygopterans.

phytotelmata. Basins that contain water formed by arrangement of leaves in some plants.

polymorphic. *adj.* Occurring in more than one form independent of sex and age.

postclypeus. Segment of face between anteclypeus and frons.

postocular spot. Pale spot adjacent to each eye on top of head in many damselflies.

propleuron. On prothorax, the sclerite immediately dorsal to the fore coxa.

prothorax. First thoracic segment, between head and pterothorax and bearing front legs.

proximal. *adj.* Toward base.

pseudo-ovipositor. Female subgenital plate that is extended and modified into a functional ovipositor.

pterostigma (also stigma). Somewhat thickened and colored cell on anterior edge of each wing near tip.

pterothorax. Main part of thorax bearing wings; fusion of mesothorax and metathorax.

quadrangle. Quadrangular cell in damselflies adjacent to arculus.

radial planate. Distinctively shaped row of cells toward wingtip.

radius. The third of the major longitudinal veins in the wing, running from base to tip.

sclerite. Individual segment of exoskeleton.

spiracle. Respiratory opening low on each side of thorax.

stigma. See **pterostigma**.

striped. *adj.* With markings that run parallel to the thoracic sutures or along the abdomen.

subcosta. The major longitudinal vein behind the costa, running from base to nodus.

subgenital plate. Plate in various forms extending backwards from front of underside of segment 8 in females; important in holding eggs when they are being laid.

suture. Connecting point between two body segments.

tarsus (pl. tarsi). Terminal, jointed part of leg, usually with 4 segments and paired claws.

teneral. Just emerged odonate, soft and pale colored.

terminal. *adj.* At end.

thorax. Middle part of insect body that contains wings and legs.

tibia (pl. tibiae). Third segment of leg, long and usually more slender than femur.

torus (pl. tori). One of pair of usually flat plates on either side of posterior end of segment 10 in dancers (*Argia*).

truncate. *adj.* Cut off sharply, with square end.

type locality. Location where type specimen collected, the specimen used in the original description and designated as typical of the species.

ventral. *adj.* Bottom or underside.

ventrolateral. *adj.* Lower side.

vertex. Area between eyes and behind ocelli, usually elevated.

Bibliography

Bailowitz, R., D. Danforth, and S. Upson. 2015. *A Field Guide to the Damselflies & Dragonflies of Arizona and Sonora*. Nova Granada Publications, Tucson.

Bota-Sierra, C. A., J. Sandoval-H, D. Ayala-Sánchez, & R. Novelo-Gutiérrez. 2019. *Libélulas de la Cordillera Occidental Colombiana, una mirada desde el Tatamá/Dragonflies of the Colombian Cordillera Occidental, a look from Tatamá*. Panamericana Formas e Impresos, Colombia. In English and Spanish.

Corbet, P. S. 1999. *Dragonflies: Behavior and Ecology of the Odonata*. Cornell University Press, Ithaca.

Esquivel, C. 2006. *Dragonflies and Damselflies of Middle America and the Caribbean*. Editorial INBio, San José. In English and Spanish.

Förster, S. 2001. *The Dragonflies of Central America, exclusive of Mexico and the West Indies: A Guide to Their Identification*, Second Edition. Odonatological Monographs 2. Gunnar Rehfeldt, Braunschweig.

Garrison, R. W., N. von Ellenrieder, & J. A. Louton. 2006. *Dragonfly Genera of the New World*. Johns Hopkins University Press, Baltimore.

Garrison, R. W., N. von Ellenrieder, & J. A. Louton. 2010. *Damselfly Genera of the New World*. Johns Hopkins University Press, Baltimore.

Kompier, T. 2015. *A Guide to the Dragonflies and Damselflies of the Serra dos Orgaos, South-eastern Brazil*. REGUA Publications (printed by Peeters NV, Leuven).

Michalski, J. 2015. *The Dragonflies & Damselflies of Trinidad & Tobago*. Kanduanum Books, Morristown, NJ.

Paulson, D. 2009. *Dragonflies and Damselflies of the West*. Princeton University Press, Princeton.

Paulson, D. 2011. *Dragonflies and Damselflies of the East*. Princeton University Press, Princeton.

Paulson, D. 2018. *The Odonata of North America, including Mexico, Central America and the West Indies*. Bulletin of American Odonatology 12(4): 35-46.

Schneeweihs, S., W. Huber, & A. Weissenhofer, eds. 2009. *Dragonflies of the Piedras Blancas National Park, Costa Rica*. Verein zur Förderung der Tropenstation La Gamba, Vienna.

von Ellenrieder, N., & R. W. Garrison. 2007. *Dragonflies of the Yungas (Odonata). A Field Guide to the Species from Argentina*. Pensoft, Sofia. In English and Spanish.

Websites

Citizen-science programs with records of Costa Rican odonates

iNaturalist (https://iNaturalist.org)
Observation (https://observation.org)
Odonata Central (https://odonatacentral.org)

Photo Credits

All black-and-white illustrations are by William A. Haber. For photos: B - bottom; L - left; M - middle; R - right; T - top.

John Abbott (44B, 95B, 148, 153, 351), Christian Aden (186B), Alfonso Auerbach (243, 257B, 327B), Giff Beaton (56B, 188B), Steve Bird (8e, 8f, 190T), Cornelio Bota-Sierra (25R, 62B, 139B, 340T), Allan Brandon (201T), David Broek (324B), Terry Carr (176T), Steve Collins (350B), Stephen Cresswell (137B, 339), Doug Danforth (207B, 212B, 213, 288T), Steven Daniel (224T, 225B), Jan Dauphin (193L), Carel de Haseth (4, 352B), Pierre Deviche (27T, 58BR, 144T, 203B), Marion Dobbs (355B), Steven Easley (203T), Cameron Eckert (57B, 82B, 87B), Joshua Emm (180T), Leslie Flint (318B), Lev Frid (78), Marla Garrison (181, 328T), Gabriel Camilo Jaramillo Giraldo (288B), Benoît Guillon (50B, 267B, 349B), William Haber (2, 10, 11, 12, 13, 16, 24, 25L, 27B, 28, 34, 37, 38, 39L, 40, 41, 42, 45, 46B, 51B, 61, 64B, 65M, 66R, 68, 69B, 72, 73T, 74, 79, 82T, 83B, 84, 89, 91T, 94, 97, 98, 99B, 100, 101B, 102, 104, 105, 109, 110B, 112, 113B, 115T, 117T, 119B, 120, 121B, 125B, 126B, 128B, 129, 130, 131T, 132T, 135B, 139T, 143 top middle, 144B, 146L, 147, 150B, 154B, 155B, 157L, 159R, 162, 165, 167, 169, 170, 171B, 175T, 178B, 179T, 189, 191L, 192, 195, 200, 202T, 204, 205, 207T, 208, 209, 210, 211B, 215, 216, 217, 219, 222, 223, 224B, 225T, 226B, 227, 228, 233, 234T, 235, 237, 241T, 244, 257T, 272T, 282B, 289T, 290T, 292B, 293, 299T, 300B, 306B, 328B, 333B, 335B, 340B, 343, 345T), James Holden (214), Hans Holbrook (345B), Rich Hoyer (172R, 250T), Wiliam Hull (234B, 251T), Eric Isley (196B), Jim Johnson (134, 143T, 146R, 250B, 251B, 304T), Justin Jones (161), Anneke Jonker (191R, 252T), Karen Kearney (135T), Tom Kompier (136B, 317B, 334, 348T, 353B), Gernot Kunz (65 inset, 75T, 156, 163, 174, 180B), Richard Kunz (124B), Greg Lasley (143 bottom middle, 149B, 193R, 268T, 341), Josh Lincoln (152), Keith McCandless (198T), Juan Carlos Garcia Morales (299B), Erland Nielsen (325, 330B, 335T, 337B, 348B), Kenji Nishida (xii, 21, 52, 76B, 239), Richard Orr (179B), Dennis Paulson (ii, 5BL, 6, 8a, 8d, 8g, 14L, 26, 29, 31, 32, 33, 39T, 44T, 55B, 56T, 58TL, 64T, 65T, 65B, 67L, 75B, 90B, 92T, 95T, 96, 103B, 108T, 115B, 116B, 122, 123B, 124T, 128T, 131B, 138, 149T, 151, 155T, 158R, 160, 166B, 168, 172L, 175B, 176B, 177, 178T, 182, 184T, 185B, 194T, 197, 198B, 199, 201B, 211T, 221T, 232, 241B, 242T, 246, 247B, 248T, 249, 260B, 261B, 262, 263T, 264, 265T, 266, 274, 275, 276, 281, 283B, 285, 289B, 290B, 292T, 294B, 295, 300T, 302, 305T, 307T, 308B, 309B, 313, 314T, 318T, 321B, 330T, 332, 333T, 336, 352T, 354, 355T, 356B, 357), Andy Pearce (186T), Ruy Penalva (185T), Liam Ragan (252B), Chris Rasmussen (85), Mike Rickard (187), Eric Roeland (283T), Jareth Román-Heracleo (267T), John Rosford (268B), Anthony Schoch (58BR), Tom Schultz (158L), Dave Smallshire (238, 326R), Netta Smith (vi, 5TL, 5TR, 5BR, 8b, 8c, 8h, 14R, 19, 22, 30, 50T, 51T, 53T, 54, 55T, 58TR, 62T, 63, 66L, 67R, 69T, 70, 71, 73B, 76T, 83T, 86, 87T, 88, 90T, 91B, 92B, 93, 99T, 101T, 103T, 106, 107, 108B, 110T, 111, 113T, 114, 116T, 117B, 118, 119T, 121T, 123T, 125T, 126T, 127, 133, 136T, 137T, 140, 142, 143B, 145, 150T, 154T, 157R, 159L, 171T, 173, 183, 184B, 188T, 196T, 202B, 212T, 220, 221B, 226T, 242B, 245, 247T, 248B, 253, 254, 255, 256, 258, 259, 261T, 263B, 265B, 269, 270, 271, 272B, 273, 277, 278, 279, 280, 282T, 284, 286, 287, 291, 294T, 296, 297, 298, 302M, 303, 304B, 305B, 306T, 307B, 308T, 309T, 310, 311T, 312, 314B, 315, 317T, 319, 321T, 322, 323, 324T, 327T, 329, 337T, 338, 342, 346, 347, 349T, 350T, 353T, 356T), Brian Steger (190B), Ken Tennessen (320), Vaughan A. Turland (194B), Yesenia Vega-Sánchez (57T), Rob Williams (166T, 260T), Wade Worthen (311B), Karen Yukich (326L).

Index to Species and Families

About the Authors

Dennis Paulson grew up in Miami, exposed to subtropical nature in all its glory while southern Florida was still largely unspoiled. He received his Ph.D. in Zoology from the University of Miami in 1966 with a study of the dragonflies of southern Florida, and shortly thereafter he moved to Seattle, where he has lived ever since. He retired as the Director of the Slater Museum of Natural History at the University of Puget Sound, where he also taught in the Biology Department. Dennis has taught at three universities and continues to teach adult-education courses in many venues. He has also led nature tours and traveled on his own to all continents, and he has studied and photographed dragonflies and birds worldwide and published over 75 scientific papers and three books on his favorite animals. He is an avid nature photographer, with many photos published in magazines, books and interpretive displays.

Dennis's published books include *Dragonflies and Damselflies of The West*; *Dragonflies and Damselflies of the East; Dragonflies & Damselflies; A Natural History; Shorebirds of North America: The Photographic Guide*; *Shorebirds of the Pacific Northwest*; *Exotic Birds*; and *Alaska:The Ecotravellers' Wildlife Guide*. Beyond these interests, he is a well-rounded naturalist with a broad knowledge of plants and animals of the world.

William Haber grew up in Wisconsin, where he developed an early interest in natural history. He has an M.S. in entomology from the University of Puerto Rico and a Ph.D. in ecology from the University of Minnesota. Bill is a resident of Monteverde, Costa Rica, where he has been involved in research, teaching, and conservation since 1972. He carried out a plant inventory of the Monteverde cloud forest with the Missouri Botanical Garden, resulting in the discovery of more than 150 new plant species. His ongoing interests include the ecology and conservation of tropical insects, especially the seasonal migration of butterflies and dragonflies and the effects of climate change on the distribution of tropical insects and plants. In 2004, Bill began an inventory and online identification guide to the dragonflies and damselflies of Costa Rica, and later also of Ecuador, where he carried out ten collecting expeditions. He has described six new species of Odonata from Costa Rica. He has published a book and authored 9 book chapters and more than 50 scientific articles. Over more than 30 years, Bill has shared his passion for the natural world with his wife, botanist and author, Willow Zuchowski. Since retiring, he and Willow have traveled widely to natural history destinations throughout the world.

Notes

Notes

Notes